Paulo Tavares Muniz Filho

Complex Hydrographic Systems

Paulo Tavares Muniz Filho

Complex Hydrographic Systems

Socio-environmental Dimensions of the Tejipió River Basin, Recife - PE Brazil

ScienciaScripts

Imprint

Any brand names and product names mentioned in this book are subject to trademark, brand or patent protection and are trademarks or registered trademarks of their respective holders. The use of brand names, product names, common names, trade names, product descriptions etc. even without a particular marking in this work is in no way to be construed to mean that such names may be regarded as unrestricted in respect of trademark and brand protection legislation and could thus be used by anyone.

Cover image: www.ingimage.com

This book is a translation from the original published under ISBN 978-613-9-61542-1.

Publisher:
Sciencia Scripts
is a trademark of
Dodo Books Indian Ocean Ltd. and OmniScriptum S.R.L publishing group

120 High Road, East Finchley, London, N2 9ED, United Kingdom
Str. Armeneasca 28/1, office 1, Chisinau MD-2012, Republic of Moldova, Europe
Printed at: see last page
ISBN: 978-620-7-86215-3

Table of contents:

SUMMARY

In recent decades, there has been an increase in the number of works dedicated to the analysis of river basins as territorial planning units, as more and more authors consider the drainage network to be the element most sensitive to the alterations caused by anthropogenic actions on natural systems. In urban centers, the systems (natural and/or social) are complex, non-linear and far from equilibrium, meaning that the responses (output) to changes in the input are always exponential, and are therefore unpredictable due to the different reactions of their constituent elements. Following this line of reasoning, this work is dedicated to the study of the Tejipió river basin as a complex system, working on its socio-spatial dynamics (components of the anthroposociological set) in a way that is not dissociated from the natural dynamics (river basin set), with the aim of drawing up a representative picture of the socio-environmental dimensions of the Tejipió river basin based on the main changes that have occurred in it over the last 50 years. This period has seen the population of the city of Recife grow and the consequent expansion of the urban sprawl over the rural areas and forests of the basin. As a result of the differential value of urban land, there was also an increase in the occupation of mangroves, wetlands and riverside areas by the low-income population, with the intensification of the practice of landfills, generally low and poorly made, which later aggravated the drainage problems, naturally present in the object of study due to its characteristics. In its 93.2 km^2 , the Tejipió river basin contains different socio-economic and cultural realities, making it a mosaic representative of Recife's social diversity.
Keywords: Systems, Tejipió River Basin, Socio-environmental dimensions, Recife.

INTRODUCTION

The complexity of large urban centers offers the researcher a challenge when it comes to reading, interpreting and understanding them. The permanent challenge of sharpening the vision to understand urban reality through its dynamic manifestations on different scales makes the city the sphinx of our time.

The search to elucidate the logic that permeates the interrelationships, the exchanges, between fixed elements, flows and society, and how the exchanges between society and the environment it constantly transforms take place, is one of the tasks of the geographer, who is involved in the difficult task of explaining the different relationships between society and nature in the elaboration of the geographical space based on the modification of the natural environment by human work. And it is precisely in urban centers that the changes caused by human activities occur most intensely, thus preventing studies dedicated to environmental impacts that dissociate them from the dynamics of the environment in which they are located.

Therefore, when researching urban environmental impacts, in addition to focusing on socio-spatial dynamics, the geographer must also pay attention to physical dynamics; and thus seek to understand how socio-spatial dynamics interfere with, alter, physical dynamics, through the intensification of pre-existing physical processes or their triggering, and how the latter condition the former, even interfering in the differential valuation of space. In other words, an integrated study should be carried out that values their inseparability.

Throughout the history of geographic thought, there are often geographers who recognize, deny or justify the study of the relationships established between man and the environment as an analysis pertinent to the science of geography; to this end, they develop work based on methodologies aimed at understanding this interrelationship, making apparent the complexity of the connections established.

The study of this type of relationship is linked to the very genesis of Geography as a science during the 19th century[1] . According to GOMES (1997: 27), "the study of the relationships between Men and Nature (in its diversity) (...), has as its starting point for Geography, the works of its classic precursors Humboldt and Ritter", who, based on a holistic vision and the scientific rigor of their observations, considered Nature as a totality, differentiated by its units or parts and by the exchange relationships they develop[2] . Carl Ritter's work deserves to be highlighted, as it was fundamental to the systematic development of Physical Anthropogeography, which Ratzel later developed through his historical understanding of space[3] .

However, with the reduction in Geography of the influence of the German School, at the time (late 19th century) accused of being deterministic because its studies were dedicated to human behavior conditioned by physical-natural factors[4] , and with the rise of the influence of the French School, which, supported mainly by the ideas of Vidal de La Blache, gave greater prominence to "socio-cultural aspects as a way of backing up geographical reasoning", the ideas of Vidal de La Blache, gave greater prominence to "socio-cultural aspects as a way of supporting geographical reasoning"[5] , the dualism between Physical Geography and Human Geography was established and, consequently, there was a reduction in works that considered the study of the relationship between Man and Nature to be essential; with emphasis now being placed on one or other aspect.

---

[1] However, attention should be drawn to the fact that "until the turn of the [19th] century, natural-scientific research dominated geography. The social question was only evoked at the beginning of this century through the work of Ratzel, although it is demonstrable, since Humboldt, that the relationship between Man and Nature is a subject of central interest to Geography" (GOMES: 1997, 30).

[2] GOMES, Edvánia Torres Aguiar. *Landscape Cut-outs in the City of Recife: a geographical approach.* Doctoral thesis. Sao Paulo: USP/CDG, 1997. p. 27.

[3] GOMES, Edvánia Torres Aguiar. Op. cit. p. 27.

[4] Despite the evident decline in the influence of the German School on Brazilian geography, it continues to have a strong influence on geographical studies developed in the USA and Great Britain.

[5] CONTI, José Bueno. *Physical Geography and Society-Nature Relations in the Tropical World.* In: CARLOS, Ana Fani Alessandri. *New Paths of Geography.* São Paulo: Contexto, 1999. p. 11.

In the 1950s, with the contribution of General Systems Theory to Geography, based on the climatic and hydrographic studies developed by Arthur N. Strahler, new perspectives opened up for integrated studies and for this science.

In the 1960s, Sotchava and Bertrand separately developed the notion of geosystems which, according to CONTI (1999:11), "gave unity and coherence to physical geography by incorporating anthropogenic action, ecological potential and biological exploitation [into analysis], while at the same time helping to blur the artificial boundaries between it and human geography".

In recent decades, as interest in environmental issues has grown, the number of works in Geography based on the socio-environmental aspect has multiplied; in this sense, "the term *socio* appears, then, linked to the term *environmental,* to emphasize the necessary involvement of society as a subject, an element, a fundamental part of the processes related to contemporary environmental problems"[6] .

This study, therefore, can be defined as developed from a socio-environmental perspective, since it aims to analyze the changes in the socio-spatial dynamics developed in the Tejipió river basin, especially in the last 50 years.

The Tejipió river basin, one of the small coastal basins in the state of Pernambuco, is a hydrographic complex made up of the Tejipió, Jiquiá and Jordao rivers and other smaller watercourses. It is located entirely within the Recife Metropolitan Region (RMR), with the source of the main river located in the municipality of Sao Lourengo da Mata and its mouth in Recife. The Tejipió River is 20.5 km long and its basin has a total area of 93.2 km^2 , encompassing part of the municipalities of Sao Lourengo, Jaboatao dos Guararapes and the city of Recife, which has the largest share (73%).

However, the dissertation focuses on the urban areas located in the city of Recife, relating them to the transformations that have taken place in the environment, with the aim of building a representative picture of the socio-environmental dimensions based on the structuring and restructuring of society in the Tejipió river basin.

In order to achieve this, an analysis was first made of the geological-geomorphological processes that took place in the area that today corresponds to the Tejipió river basin, with a view to understanding the formation of the different physical units that make up the basin. This is followed by a review of the historical process of urban occupation of the lakes contained in the basin, highlighting the role played by networks in the formation of neighborhoods and the distribution of the population over them. Finally, the relationship established between the agents that shape the area and the natural setting modified by their actions is understood.

As the dissertation aims to understand the different relationships between the anthropic and physical constituents through the analysis of socio-environmental characteristics, it was decided to develop it based on a systematic conception of reality, capable of providing a view of the watershed as a whole, fragmented and, at the same time, articulated from the interrelationships of its parts.

By providing an understanding of reality based on the interrelationships between the component parts of the analyzed whole and the action of this whole on its parts, the systematic method promotes knowledge of geosystems based on their internal and external dynamics[7] . It also provides knowledge of the concrete object, based on the relationships it maintains with the other objects that make up its environment, which allows us, in addition to knowing it, to define it.

The work acquires relevance in the application of the systematic method to the study of a watershed located in an urban environment, within which, as it is a complex, non-linear and far from equilibrium system, the responses will always be given in an exponential way, making it difficult to predict the phenomena and the responses given by the system to variations in the energy input.

---

[6] ESTEVES, Cláudio Jesus de Oliveira. *Tourism and Water Quality on Ilha do Mel (Paraná Coast)*. Master's dissertation. Curitiba: CPGG, 2004.

[7] Pascal, still in the 17th century, was already drawing attention to the fact that it is impossible to know a whole without knowing its constituent parts and, conversely, that it is impossible to know the parts without knowing the whole of which they are a part; he expressed this in his book entitled *Pensées*: "I consider it impossible to know the parts without knowing the whole, as well as to know the whole without knowing the particular parts."

As previously mentioned, the application of Systems Theory to the study of river basins is not recent, with the first studies of this nature in Geography dating back to the 1950s. However, the majority of these studies are concerned with rivers located in rural or sparsely urbanized areas and work by dissociating anthropic dynamics from physical dynamics.

The aim here is to work on these two dynamics in an integrated way, based on an understanding of the Tejipió river basin as a complex system, in which its reality is constructed from the interrelationship between the river basin itself (the set of lands drained by the Tejipió river and its tributaries) and the socio-spatial dynamics developed over it. Changes in environmental dynamics will be reflected in changes in social dynamics and changes in social dynamics will promote changes in physical dynamics, due to the destabilization of natural systems.

On the other hand, the study acquires relevance because it deals with a river basin that is not analyzed in the surveys carried out by the state government, nor in those carried out by Recife City Hall, and is not even included in the division of the state into river basins (which from our point of view, explained in subchapter 3.1, should be included in the Group of Small Coastal Basins - II [GL-II]). However, in the current political-administrative scenario, with the resumption at state level of discussions on ideal forms of water resource management, corroborated by the reforms implemented by the state government, such as the creation of the CRH (State Water Resources Council), work that discusses environmental impacts linked to socio-economic activities developed in river basins becomes pertinent.

CHAPTER 1

In the intricacies of the theoretical elements and methods of approach to analyzing the problem.

1.1 From Systematization to Developments - revisiting General Systems Theory

I consider it impossible to know the parts without knowing the whole, as well as knowing the whole without knowing the particular parts"
Pascal

The aim of this study is to understand and explain the different relationships that exist between the anthropic and physical constituents that co-form the territorial configuration of the Tejipió river basin. In order to achieve this goal, an analysis is made of the socio-environmental characteristics and metamorphoses[8] that have taken place in this area over the last 50 years. This chapter sets out the theoretical frameworks and research methods identified as suitable for analyzing the problem. It starts with the theoretical conceptualization of the River Basin phenomenon, going through its complexity and the articulations engendered in and by it.

The hydrographic basin, as a discrete unit in the complex context of the earth's surface, will be understood here as: the area drained by a main river, its tributaries and sub-tributaries, within which a hierarchical ordering of the watercourses can be established, according to the ordering procedure developed by Strahler. Its area is defined by the portion of space drained by the whole river system (drainage network) with a common mouth in a delta or estuary, as is the case with the Tejipió River.

The term complex comes from the Latin *complexus* and means "woven together"[9]. So, starting from the point of view of the coexistence and simultaneity of events[10], both physical and socio-economic-spatial, the temporal succession of different uses[11], we define the area drained by the Tejipió river basin as a complex system; a space in which socio-spatial and physical dynamics are no longer explained by the constituent parts, nor by analyzing the whole as an irreducible whole, but rather by the interrelationships, the exchanges, between this whole and its parts.

The word "system" has an old and widespread use in scientific knowledge and is generally used to designate an organized set of elements and interactions between them, such as the solar system. The system appeared in science as a supporting concept, without any elucidation of the concept of system, until the 1950s, when it came to be considered as an entity, not as an agglomeration of parts[12]. In its general sense, MORIN (1997) considers the term system to be an enveloping word and that in "its particular sense it adheres totally to the matter that constitutes it: it is therefore impossible to conceive of any relationship between the various uses of the word 'system': solar system, atomic system, social system; the heterogeneity of the constituents and principles of organization between stellar systems and social systems is so evident and impressive that it annihilates any possibility of uniting the two meanings of the term 'system'".[13]

The systems approach provides a dynamic view of the interactive processes developed by physical and anthropic agents within the system in question. Aware of the complexity reached by the interrelationships developed in environmental systems, particularly urban environmental systems, in which the responses will not always be given in proportion to the modifications evidenced in the input, i.e. in a linear way, and in which the processes in the elaboration of forms will be altered by

---

[8] The metamorphoses are the result of changes in the territorial configuration of the basin, promoted by changes in production techniques and methods, through the materialization of new events.

[9] MORIN, Edgar. *The Method.* Lisbon: Publicacoes Europa-América, 1997.

[10] SANTOS, Milton. *The Nature of Space: Technique and Time, Reason and Emotion.* Sao Paulo: EDUSP, 2002.

[11] MONTEIRO, Carlos Augusto de Figueiredo. Geosistemas: the history of a search. Sao Paulo: Contexto, 2000.

[12] Although some conceptions of a system as a set of parts forming an organized global unit date back to the 17th century (Leibniz considered a system to be a set of parts), they are still incipient, diffuse and poorly structured within a body of theory.

[13] MORIN, Edgar. Op. cit. p.98.

them, In a permanent dialectic between products and conditioning factors, dynamics and results, we opted to develop the study supported by a systematic conception of reality, capable of providing a view of the Tejipió watershed as a whole, fragmented and, at the same time, articulated from the interrelationships of its parts.

The General Systems Theory, developed in the 1950s by Ludwig von Bertalanffy, published in English in 1968 and in Brazil in 1973, was the result of the concern of the Theoretical Biology of the 1930s to carry out a conceptual and analytical systematic approach[14] of living organisms, considered as an organized totality, which led to a change in scientific methods with the introduction of a new principle: holism[15] . Because of the organicism that exists in many fields of scientific knowledge and its practical applicability to the most diverse objects of study, the systematic approach was absorbed and adapted to various other disciplines. Until the 1980s, biology remained the main theoretical influence on systemic studies, but since then, due to the advances made in studies in physics and chemistry throughout the 1970s (systems theory), the systemic approach has been adapted to other disciplines.
Dynamics, Chaos Theory), the latter have supplanted the perspectives related to Theoretical Biology.

In formulating the general systems theory Bertalanffy hoped to elaborate, on the basis of his considerations, the postulate of a new discipline whose "object is the formulation of principles valid for 'systems' in general, whatever the nature of the elements that compose them and the relationships or 'forces' existing between them"[16] , thus considering the general systems theory the general science of totality with the purpose of integrating the various sciences, natural and social, under the same postulate[17] .

Systemic theory is based on a holistic approach[18] to reality (some authors even use general systems theory as a synonym for holism, e.g. CHRISTOFOLETTI, 2004). This approach considers it possible to understand a totality only by studying it as a discrete whole within the whole of nature. The holistic approach replaces the analytical approach in the sciences as a whole, although in some of them, such as geography, one is not antagonistic to the other, and they can even be considered complementary.

The analytical approach, also known as reductive, predominated in the scientific field throughout the 19th century, exercised within the framework of what is known as Cartesian science[19] . In this approach, the whole is broken down in favor of analyzing its constituent parts, considering that knowledge of how the parts work leads to knowledge of the whole. Thus, the problem is focused

---

[14] CHRISTOFOLETTI, Antonio. *Modeling Environmental Systems*. Sao Paulo: Edgard Blücher, 1999. p.5.

[15] However, we recognize that it is not possible to attribute the origin of systems theory solely to the theoretical concerns of Theoretical Biology. The constant advances made in technology throughout the 20th century led to changes in the way objects were perceived; we could no longer think in terms of isolated machines, but in terms of "systems". The complexity presented by the new inventions escaped the competence of a specific field of knowledge, calling into question the existence of "specialists" and the watertight view of objects. Bertalanffy himself admitted this fact when he stated that "when it comes to ballistic missiles and space vehicles, these devices have to be made up of components originating from heterogeneous technologies, mechanical, electronic, chemical, etc." (1977,18); unlike a steam engine, a radio or even an automobile, in which the problems of operation and/or manufacture can be solved from a specific field of engineering. He goes on to admit the limits of simple quantification when he states that "air traffic or even car traffic is no longer a question of the number of vehicles in operation, but form systems that must be planned and organized" (1977, 18). We are afraid to say that the formulation of General Systems Theory is imbued with the "Spirit of the Times" ("*der Zeitgeist*").

[16] BERTALANFFY, Ludwig von. *General Systems Theory*. Petrópolis: Vozes, 1977. p.61.

[17] Within the author's quantitativist vision, possibly the fruit of his neo-positivist education in Moritz Schlick's group (the so-called Vienna Circle), he also endowed his theory with the pretension of "achieving an exact theory in the non-physical fields of science" (BERTALANFFY: 1977, 62).

[18] The holistic approach analyzes the whole, inducing the conception of a whole that is greater than the sum of its parts, because in the whole there are characteristics that do not come from the sum of its constituent parts, new characteristics, emerging from the whole. Therefore, holism asserts that it is impossible to understand and study the parts by analyzing them directly, since they are coerced by the functioning of the whole.

[19] Cartesian science is based on the Cartesianism of René Descartes (1596-1650), which is characterized by rationalism and by considering a phenomenon or concept in isolation from the totality in which it appears. The Cartesian method focuses on clear ideas and rigorous procedures, and is sometimes limited to mechanistic, simplifying explanations, which are therefore inadequate for understanding reality.

on the lower level of the hierarchy of complexity, despite the fact that the history of knowledge has proven "that the behavior of nature and man is more complex than the simple application of the reductionist method"[20] In the reductionist approach "every object can be defined on the basis of the general laws to which it is subject and the elementary units of which it is made up"[21] .

However, MORIN (1999) points out that holism depends on the same simplifying principle as reductionism, despite the apparent antagonism between the two, because it exposes a simplified idea of the whole and a reduction of the whole to itself. The aforementioned author states that by considering the system a global unit, holism merely replaces the simple elementary unit of reductionism with a simple macro-unit, as it does not consider the intrinsic qualities of the parts that are not reducible to the whole[22] . Taking atomic, biological and social systems as examples, he says that "a system is not only a constitution of unity from diversity, but also a constitution of (internal) diversity from unity"[23] , because in the context of totality there is the emergence of new qualities at the level of the parts independent of the control and coercion of the whole (not only does the individual ignore and is unconscious of the social totality, but the latter also ignores and is unconscious of the individual's most intimate aspirations); modifications occurring at the level of the parts in natural systems can be amplified through the mechanism of feedback until, in the end, they modify the functioning of the whole).

Despite its generality, the concept of a system finds applicability in the most diverse fields, implementing considerable changes by replacing the notion of an object[24] with the notion of a system: a complex whole formed from the interaction between its forming elements and greater than the sum of its characteristics, whose functioning is affected by the environment and which cannot be understood outside this context, in isolation. What was an object in classical science is now considered a system: atom, society, watershed, city, region.

In Geography, the systematic methodology was introduced in the 1950s through the geomorphological and climatic studies carried out by Arthur N. Strahler. With the advent of theoretical-quantitative geography, it became one of the main theoretical-methodological tools in geographic studies due to its applicability to planning.

CHRISTOFOLETTI (1979), in the compendium he organized, carries out an analysis of the application of systems theory to Geography and, to this end, reproduces some definitions of systems. In all of them, the need for the existence of a set of elements (or units) that maintain relationships with each other is evident for the constitution and existence of a system. From these considerations we can conclude that systems are a set of elements or units organized by virtue of the interrelationships between them.

However, when differentiating between two of the definitions he worked on, CHRISTOFOLETTI highlights the functional nature of one of them, stating that "systems are organized to achieve a certain purpose in the whole of nature"[25] , exemplifying the case of a watershed, explaining that it was organized to drain water and debris provided by the drainage process. If we accept this, we will be endowing nature with intentionality. Watersheds drain water and debris, but they are not organized for this purpose. Their organization depends on physical randomness, on the arrangement of their forming elements in the space occupied by this system and, above all, on the anthropic practices that animate them.

MORIN (1997 & 1999) gives the interrelationships between elements a central role in the formation and understanding of systems, since organization, the notion that gives systems their

---

[20] CHRISTOFOLETTI, Anderson Luís Hebling. *Dynamic Systems: The approaches of Chaos Theory and Fractal Geometry in Geography*. In: VITTE, Antonio Carlos & GUERRA, Antonio José Teixeira (org.). *Reflexoes sobre a Geografia Física no Brasil*. Rio de Janeiro: Bertrand Brasil, 2004. p.90.

[21] MORIN, Edgar. Op. cit. p.94.

[22] "The whole is indeed a macro-unit, but the parts are not merged or confused in it; they have a double identity, their own identity which remains (...) a common identity, that of their systemic citizenship" (MORIN: 1999, 260).

[23] MORIN, Edgar. *Science with conscience*. Rio de Janeiro: Bertrand Brasil, 1999. p.260.

[24] A closed and distinct entity that would be defined in isolation in terms of its existence, its characteristics and its properties, independent of the environment.

[25] CHRISTOFOLETTI, Antonio. *Systems Analysis in Geography*. Sao Paulo: HUCITEC, 1979. p. 3.

phenomenal existence, comes from the interaction between their constituent parts or, in his words, "in a system, the interrelationships between elements/events or individuals are constitutive of the totality and, therefore, constitute the organization of the system"[26]. With this, he distances himself from the formulations of CHRISTOFOLETTI (1979), who linked the concept of system to the functional perspective, highlighting the functionality of systems as the basic standard for characterizing them.

According to Milton Santos, when we analyze space taking into account only the elements, the nature of the elements or their possible classes, in isolation, we have not gone beyond the realm of abstraction. In this way, SANTOS' (1997) reasoning converges with MORIN's, when the former considers that we can only know and define an object based on the relationships it maintains with other objects, since it does not have an isolated existence. Quoting KOSIK (1967), he points out that "the interdependence and mediation of the part and the whole mean, at the same time, that isolated facts are abstract elements artificially separated from the whole and that only through their participation in the corresponding whole do they acquire veracity and concreteness. In the same way, a whole in which the elements are not differentiated and determined is an empty whole."[27].

This introduces the question of method, which points to the irreducibility of the whole to the parts and of the parts to the whole, with the whole being greater than the sum of the parts and the parts, at the same time, being greater and smaller than the whole, due to the individual emergencies that arise at the level of the parts, independent of the control of the totality.

CHRISTOFOLETTI (1979) emphasizes that in order to understand geosystems, they need to be seen as "natural systems? that "depending on the individual properties of the subsystems, the same external influence can have different consequences"[28]. This fact rules out the establishment of a rigid relationship between cause and effect in natural environments, on which some studies on environmental impacts are based, particularly urban environmental impacts. When studying the urban environment, it must be considered as a complex, non-linear system, far from equilibrium, in which responses will always be given exponentially to new inputs into the system[29].

This dynamic aspect attributed to natural systems is in line with the formulations of COELHO (2001), for whom the problem of urban environmental impacts[30] must be dealt with as a dynamic process, as "a process in permanent movement" which is "at the same time a product and producer of new impacts."[31] The author believes that the study of urban environmental impacts is directly linked to interpretative scales, whether spatial or temporal. The pollution of the Tejipió River, for example, is linked to multiple causes, which are temporally and spatially diverse, although they are interconnected with processes that have both local causes and broader ones that vary over time. This does not exclude the fact that local causes may be predominantly responsible for a phenomenon (pollution linked to industrial concentration in a particular part of the geographical space, for example).

---

[26]MORIN, Edgar. Op. cit. p. 99.

[27] KOSIK, Karel apud SANTOS, Milton. *Space & Method*. São Paulo: Nobel, 1997. p.14.

[28] CHRISTOFOLETTI, Antonio. Op. cit. p. 12-13.

[29] CORREA, Antonio Carlos de Barros: 2004. Class notes.

[30]When analyzing the environmental impact of cities, COELHO (2001) uses political economy or the political ecology of the environment, based on the ideas of Marx and Engels. Political economy claims to encompass in its analysis the imbrications between ecological, political-economic-spatial and socio-cultural processes. In short, it aims to examine "... the dynamic relationships between nature and society and temporally determined socio-spatial structures" (2001:26). These temporally determined socio-spatial structures are the result of the intersection between physical-chemical, political-economic and socio-cultural processes; a reflection of "... the way in which social classes and the economy are structured and de-structured in the face of external intervention" (2001:27). They have a temporal character because of the ruptures that occur in each of the processes, which give rise to a new, relatively stable structure until another rupture destabilizes it. In order to understand this process, it is important to bear in mind that "ruptures from different causes trigger... processes of combined ecological and social change, i.e. environmental impacts of a structural nature, which produce new changes that affect social class structures in a differentiated and unplanned way" (2001:27).

[31] COELHO, Maria Célia Nunes. *Environmental Impacts in Urban Areas - Theories, Concepts and Research Methods*. In: GUERRA, Antonio José Texeira & CUNHA, Sandra Batista da (eds.). *Urban Environmental Impacts in Brazil*. Rio de Janeiro: Bertrand Brasil, 2001. p. 25.

In this study we will understand the geographical space as a socially constructed concreteness, as a dimension of the relationship between society and nature, which, in its material and objective dimension, is a product of the transformation of nature, or natural space, by social work[32].

And we will understand environmental impacts as the process of social and ecological changes caused by disturbances in the environment[33]. In this way, the environment is socially and historically constructed, the result of constant interaction between a society on the move and a particular physical space that is permanently modified[34].

This understanding of impacts as a process will depend on a non-linear understanding of the history of their production, the urban development model and the internal patterns of social differentiation.[35]

According to CHRISTOFOLETTI (1999), environmental systems "represent organized entities on the earth's surface, so that spatiality becomes one of their inherent characteristics. The organization of these systems is linked to the structure and functioning of (and between) their elements, as well as resulting from evolutionary dynamics." Environmental systems, by virtue of their broad wording, end up having biological, physical and anthropic elements as components, which allows them to be subdivided into ecosystems and geosystems[36]

According to the Ministry of the Environment, an ecosystem is "a dynamic complex of plant, animal and microorganism communities and their inorganic environment, which interact as a functional unit". However, we consider it necessary for understanding ecosystems to include man and his dynamics as transforming agents who, by interacting with others, modify previous environmental conditions.

Similarly to geosystems, ecosystems - considered here as biomes - present themselves as discernible spatial units in the whole of nature, and boundaries can be established for them through the interaction between their various components, but taking into account the vertical flow of energy between the different species (food chain), which is responsible for maintaining and characterizing ecosystems. Thus, from an ecosystem perspective, the delimitation of ecosystems depends on the "vertical integration of the habitat and the organisms that make up the entity"[37].

The geosystem is an environmental system that arises from the dynamic interaction of physical, biological and anthropic elements, in which "the products of socio-economic systems enter as inputs and interfere in the processes of flow of matter and energy, even having repercussions on the responses of the geosystemic spatial structure"[38] The term geosystem, coined by SOTCHAVA in 1962, has as its main notion the integration of the dynamics of nature and society, "focusing on integrated aspects of natural elements in a spatial entity".[39]

As natural phenomena, although they include socio-economic factors, geosystems result from the "combination of ecological potential (geomorphology, climate, hydrology), biological exploitation (vegetation, soil, fauna) and anthropic action"[40] (figure 1), without necessarily presenting homogeneous physiognomic characteristics, but rather an essentially dynamic complex, due to the variability of the elements, presenting itself as a unique and inseparable whole.

---

[32] SOUZA apud COELHO: 2001, 23.

[33] COELHO, Maria Célia Nunes. Op. cit. p. 24.

[34] COELHO (2001) sees environmental impact not only as the result of an action on the environment, but also as a relationship between social and ecological changes in motion. In the production of environmental impacts, ecological conditions are altered by historical, social and cultural conditions, and the latter are altered by ecological conditions. So we have a dialectical dynamic in motion, where the impact is both a product and a producer of new impacts. As a product, it acts as a conditioning factor for the process that follows. Therefore, in this study we will only be able to portray one of the stages of the movement that will continue.

[35] COELHO, Maria Célia Nunes. Op. cit. p. 35.

[36] Sotchava emphasizes the fact that "there is no basis for placing the sign of equality between geosystems and ecosystems... The merging of these concepts, besides not promoting the progress of either Geography or Ecology, is incorrect" (SOTCHAVA apud MONTEIRO: 2000, 48).

[37] CHRISTOFOLETTI, Antonio. Op. cit. p.53.

[38] CHRISTOFOLETTI, Antonio. Op. cit. p.43.

[39] CHRISTOFOLETTI, Antonio. Op. cit. p.43.

[40] CHRISTOFOLETTI, Antonio. Op. cit. p.42.

FIGURE 1 SCHEMATIC FLOWCHART OF GEOSYSTEM INTERACTION

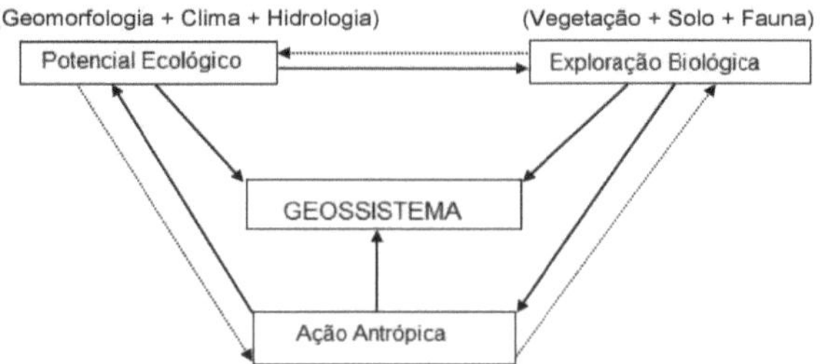

Source: Bertrand: 1968, adapted from Monteiro: 2000, 31.

In the city of Recife, as in any other large city, environmental problems (ecological and social) do not affect every urban space equally. The differential rent of urban land is a factor that will predominantly affect the physical space occupied by the less privileged classes. The distribution of these classes is associated with the devaluation of the space, whether due to its proximity to river floodbeds or to its unhealthiness, in the case of Recife caused by environmental risks such as landslides and erosion. On the other hand, it is important to note that other factors will play a part in the devaluation of the area, be they locational or socio-cultural.

1.2 Tejipió River Basin: A View of a Complex System

"... what we call constituent parts form an inseparable whole, which can only be studied together, because the part does not allow us to recognize the whole, nor should the whole be recognized in the parts..."

Goethe

Following the systemic theoretical framework and the inseparability of social and physical dynamics, we represented the Tejipió river basin as a complex system, seeking to portray the different interrelationships between the watershed (watershed set) and the socio-spatial dynamics (anthroposociological set) developed in the spaces it contains. These two groups are interconnected through the drainage network (natural, artificial and artificial) and the implications that the quality of the sanitation and waste collection networks will have on the environmental quality of the waters (figure 2).

FIGURE 2 GENERAL RELATIONSHIPS BETWEEN FACTORS INVOLVED IN THE DYNAMICS DEVELOPED IN THE TEJIPIÓ RIVER BASIN

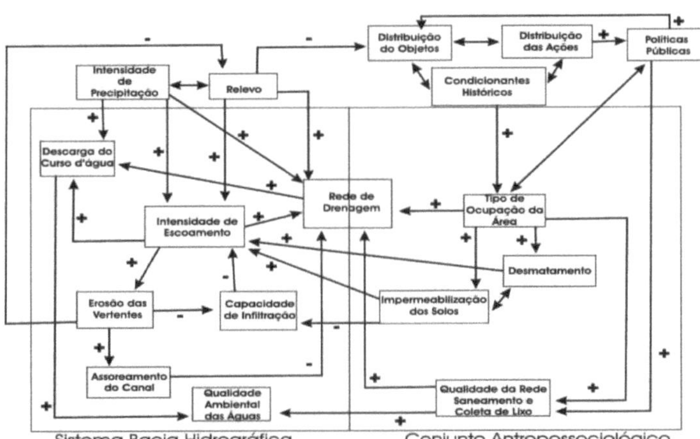

Note: the arrows suggest independence and dependence; the double arrow links the interaction of two variables; the mathematical signs indicate whether the variations are direct (+) or inverse (-).

The system represented on the previous page reproduces some of the interactions developed in the basin in question; it is made up of some of the elements we use in our analysis with the aim of elucidating how socio-spatial dynamics act to transform environmental dynamics. And in this process, the inclusion of class structure in the analysis will provide a broader view of the processes developed and their implications. Above all, we must "... investigate... the impediments, inadequacies and conflicts between natural forces and anthropic use and actions. This means highlighting the interdependencies that interconnect systemic organization."[41] .

We will now try to describe the attributes and interactions between the component elements of the systemic model proposed here. However, we will only focus on their theoretical description, without presenting empirical results. This would take more time than it would for a master's thesis. The description will be concise, since the analysis of a few interactions or even a few elements in themselves would meet the requirements for a dissertation. However, this will not detract from understanding the dynamics observed in the Tejipió river basin.

Generally, the relief is adopted as the controlling unit in geosystemic formulations. As in Bertrand, who developed his concept of geosystem at the end of the 1960s independently of Sotchava, elaborating his concept based on his studies in the French Pyrenees[42] .

A clear example of the use of relief as a control unit in a geosystem is the division of the city of Recife into geoenvironmental units, which divides the city into two distinct units: the hills and the plain.

The flowchart presented above is no exception: the relief is taken as the control unit. This is because it will be responsible for individualizing the Tejipió river basin from the Capibaribe basin and the Jaboatao basin, as well as assigning the characteristics of the drainage network.

The landforms are the result of the geological-geomorphological processes that acted during the formation of the basin (described in chapter 2.1) and, in the current period, the landforms act as conditioning factors for the processes, interfering in their distribution in the basin. Depositional processes predominate in low-lying areas, while erosive processes predominate on the slopes. However, in the analysis, the constituent material of the geomorphological features must be taken into account, as different materials can give rise to different processes, despite the shapes being the same. Geomorphological aspects also have an inverse influence on the distribution of objects, taking into account the "predilection" of events to materialize in flat areas that facilitate access, the flow of production and communication with other objects.

Relief interacts with climate, conditioning it and being conditioned by it. Due to the breadth of the term climate, which encompasses various atmospheric dynamics, in our study we adopted precipitation intensity as a relevant element for analysis.

The dissection of landforms is influenced by the variation in rainfall throughout the basin. This variation, in turn, will be influenced by the landforms that it helps to flatten; the landforms dam up the humid air masses, causing rainfall to intensify in some places due to orographic influence. This can be seen in the greater intensity of rainfall in areas close to the elevations to the west of the basin.

Precipitation intensity is the main input to the watershed geosystem. It provides the energy needed to initiate geological and geomorphological processes, as well as supplying the drainage network by increasing the discharge of watercourses and the intensity of surface runoff, giving rivers the energy needed to maintain their capacity to transport debris and to maintain erosive and depositional processes along their different stretches (upper, middle and lower reaches). However, in the case of the intensity of surface runoff, rainfall can only condition its expansion. This is linked to the infiltration capacity of the soil (+)[43] , which depends on the level of deforestation and sealing (+),

---

[41] MONTEIRO, Carlos Augusto de Figueiredo. Op cit. p.98.

[42] At the risk of sounding deterministic, MONTEIRO (2000, 47) attributes the influence of the environment to the basic differences in the formulations of Sotchava and Bertrand, who classified their geosystems on the basis of biogeographical formations and taxonomic orders of relief respectively. For the author, "it seems logical that the Frenchman, working in the Pyrenees - where the changes are accentuated in altitude - would resort to relief. Meanwhile, the Russian, working on the Siberian plains, would naturally turn to the biotic cover (vegetation-animals) as his main support."

[43] The existence of mathematical addition or subtraction signs in brackets in the body of the text will henceforth serve as an indication of the variations presented by the interaction of the elements. Addition signs (+) stand for direct variation and subtraction signs (-) for inverse variation. As shown in the flowchart on page 37.

and the steepness of the slopes (+).

When there is an increase in the intensity of surface runoff, there is also an intensification of slope erosion (+), which is responsible for silting up the river channel[44] (+) and reducing the slope of geomorphic features (-). Slope erosion can be aggravated by the reduction in soil infiltration capacity (-) caused by deforestation or soil sealing (+), which in turn contributes to the intensification of surface runoff (+).

The process observed in the previous paragraph is a typical example of negative feedback. In all the dynamics presented so far, the processes are interlinked, signaling their interdependence. However, it is only in the latter that the effects triggered by an event generate a sequence of phenomena whose consequences act once again on the initial object or fact.

CHRISTOFOLETTI (1979 & 1982) considers that "in order for there to be feedback, there must be a closed causal circuit"[45] , where any change in the first element (the one in which the process begins) is subsequently altered by the return action promoted by the feedback, after the circuit.

He points out four basic types of feedback: direct feedback, loop feedback, negative feedback and positive feedback. Below we'll go into more detail about direct and negative feedback, which are present in our flowchart[46] .

Direct feedback occurs when there is a direct relationship between two elements or variables. In our model, this type of feedback is represented by the precipitation intensity and relief variables (figure 3)[47] .

FIGURE 3 DIRECT FEEDBACK DEVELOPED IN THE TEJIPIÓ RIVER BASIN

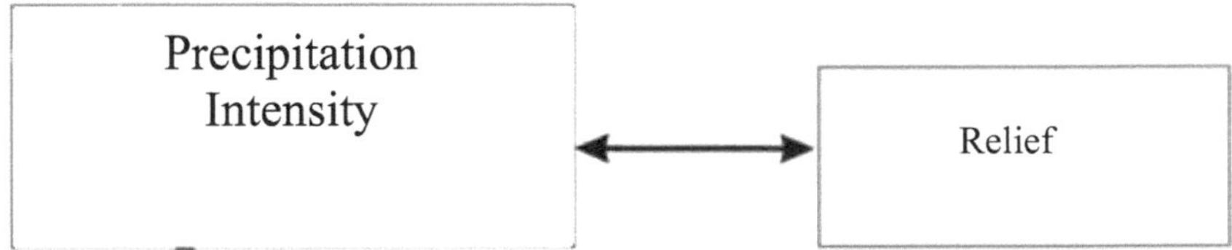

Positive feedback occurs when the circuit between the variables reinforces the action produced externally through their interaction, so that the changes are always in the direction of the initial influence. They may not have negative correlation signs, but if they do, they should always be even-numbered. In general, this type of feedback does not promote the stabilization of the system, but rather its acceleration and amplification of transformations in only one direction, and often its destruction. Normally, systems that develop positive feedback are re-stabilized before they are destroyed. Figure 4 represents a negative feedback system, the explanation of which can be found in the previous paragraphs.

FIGURE 4 POSITIVE FEEDBACK DEVELOPED IN THE TEJIPIÓ RIVER BASIN

---

[44] In addition to sediment from the normal erosion process, in basins located in urban areas, there is a significant increase in sediment due to construction (land clearing, construction of streets, avenues and highways, among others) with significant environmental consequences for the city (TUCCI: 1995, 29).

[45] CHRISTOFOLETTI, Antonio. Op. cit. p. 23.

[46] All definitions elaborated according to CHRISTOFOLETTI: 1979 & 1982.

[47] However, we believe that this type of feedback can be made up of more than two elements in a more complex way, as can be exemplified by the direct interrelationship between the distribution of objects, the distribution of actions and historical conditioning factors.

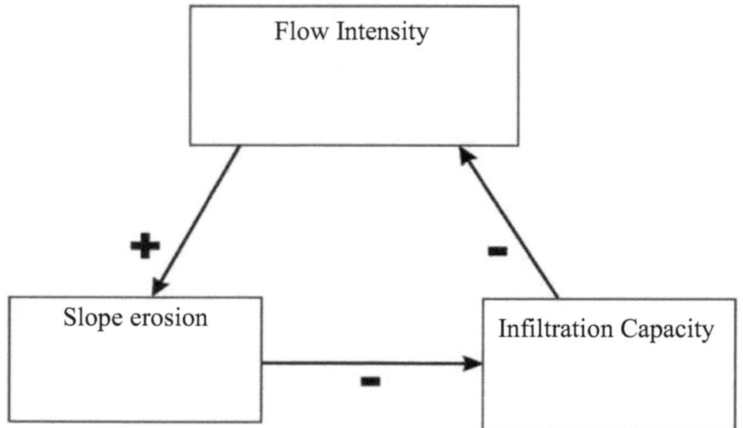

The intensification of surface runoff will have a direct impact on the drainage network, either by increasing the amount of water entering the system, or by increasing the discharge of watercourses (+), or by silting up channels through the intensification of slope erosion.

The discharge of the watercourses will have a great influence on the environmental quality of the waters, because the greater the volume of water in the rivers that make up the basin, the greater their capacity to dissolve the pollutants that are washed into them. In general, the he environmental quality of river water decreases during the spring-summer (dry period) and increases during the autumn-winter (wet period).

However, the environmental quality of water cannot be defined solely by this variation between wet and dry periods; it is directly linked to the quality of sanitation networks and waste collection.

Dumping points for domestic and industrial effluents are easily identified along the course of the Tejipió River. The former are mainly responsible for the rise in BOD (Biochemical Oxygen Demand) and FC (Fecal Coliforms) levels and the consequent reduction in DO (Dissolved Oxygen). Below is a table showing the variation in these levels according to monitoring carried out by the CPRH (Companhia Pernambucana do Meio Ambiente) between 1995 and 1998 as part of the Environmental Water Quality Program (PQA).

The organic load released into liquid bodies is measured in BOD. Biochemical oxygen demand expresses the amount of oxygen that aerobic organisms consume to simplify organic matter into substances such as CO_2, NH_3, H_2O and mineral salts[48] . Thus, when a high organic load is introduced into the environment (via domestic or industrial waste), there is an artificial demand for oxygen, which is consumed by aerobic organisms during the simplification of organic matter. Therefore, during this process, there is a reduction in the oxygen dissolved in the water (DO), which is vital for the maintenance and development of life, as well as an increase in the levels of carbon dioxide (CO_2) and a reduction in the pH of the water, consequently making it acidic[49] .

---

[48] CPRH. *Monitoring the Quality of Water in the Hydrographic Basins of the State of Pernambuco - 1998*. Recife: CPRH, 1998.

[49] BOTELHO, Rosangela Garrido Machado & SILVA, Antonio Soares da. *Watershed and Environmental Quality*. In: VITTE, Antonio Carlos & GUERRA, Antonio José Teixeira (org.). *Reflexoes sobre a Geografia Física no Brasil*. Rio de Janeiro: Bertrand Brasil, 2004. p. 178.

TABLE 1: DBO, OD AND CF INDICES OBTAINED FROM THE CPRH MONITORING BETWEEN 1995 AND 1998 - PQA

	1995			1996			1997			1998		
TABLE OF CONTENTS MÉS	BOD (5 days at 20°C) (mg/l)	OD (mg/l)	CF 100ml	BOD (5 days at 20°C) (mg/l)	OD (mg/l)	CF 100ml	BOD (5 days at 20°C) (mg/l)	OD (mg/l)	CF 100ml	BOD (5 days at 20°C) (mg/l)	OD (mg/l)	CF 100ml
January	4,30	2,80	24.000	-	-		11,40	0,00	160.000	8,80	4,00	13.000
February	4,30	4,60	22.000	-	-		6,00	11,3	35.000		-	
Marco	-			12,90	4,80	160.000	-		-	20,00	4,80	160.000
April	10,70	1,00		4,70	0,60	160.000	6,60	0,00	160.000		-	-
May	-			10,60	0,00	21.000	2,00	4,4	90.000	5,00	2,40	-
June	2,90	2,20	13.000	-	-		-		-			-
July	4,00	0,80	160.000	-	-		2,00	3,4	50.000	6,00	3,60	160.000
August	6,70	7,30	160.000	13,40	1,20	160.000	-	3,2	50.000		-	-

September	9,00	6,30	160.000	9,70	0,40	14.000	20,00	0,00	160.000	6,00	7,60	2.700
October	3,20	4,80	23.500	5,20	8,00	24.000	20,60	0,00	160.000		-	-
November	11,00	3,20	160.000	-	-		-	-	-	16,90	0,00	90.000
December	22,10	4,00		17,80	0,00	8.100	-	-	-	-	-	-
Annual Average	7,82	3,7	72.250	10,61	2,14	78.157	9,80	2,8	108.125	10,45	3,7	85.140

Source: CPRH/PQA-1995; 1996; 1997; 1998.

CF monitoring is carried out due to the nature of this bacterium, which serves as an indicator of the quality of water for human use. Bacteria of the coliform group are abundant in the feces of warm-blooded animals (fecal pollution) and are more resistant than pathogenic bacteria, so it is assumed that water which does not contain a high concentration of coliforms is not harmful to humans[50] .

As mentioned above, the quality of the waste collection network will directly affect the environmental quality of the water, but it is not enough to say that there is a strict cause and effect relationship between the two. According to EMLURB (Empresa Municipal de Limpeza Urbana), garbage collection in RPA's 5 and 6 is carried out daily, except in locations that are difficult to access or that present any kind of obstacle to collection, which is done every other day. However, despite this, in 1996 16,751 cubic meters of solid waste were removed from the Tejipió river channel; in 1999, with the expansion of the collection area from Jardim Uchóa to Totó, this figure rose to 52,272m^3 of solid waste; in 2003 the volume collected rose to 230,880m^3 , in the first seven months of that year[51] .

The data above shows that, if we take the premise of daily collection as true, the issue of garbage goes beyond the removal of garbage from the streets by municipal bodies. It also involves educational issues and raising the awareness of the poor people who live on the banks of rivers and canals and use them to dispose of their waste. As well as contributing to the silting up of water bodies, garbage is also a vector for diseases, making it a public health problem.

The distribution of objects obeys the logic of accumulation and the (re)production of capital. Distribution is linked to decisions, almost never made on the spot, and the networks superimposed on the basin that organize, establish and enable exchanges between this unit and other areas of Recife's urban space, as well as national and international locations.

However, the objects should not be thought of and analyzed in isolation from the auguries, as this would nullify their philosophical reality and fail to account for their historical reality, and the

---

[50] CPRH. Op. cit.

[51] Solid waste is not collected annually from the river gutters.

latter is also true for the system of auguries, which do not occur without the system of objects[52] .

Actions take place in space through and from the system of objects, giving them foundations that are pertinent to the intentionality of those who have the power to make decisions (governments, companies, international organizations, among others). When action is exerted on an object, it modifies it and is ultimately modified by it, and these two movements are concomitant[53] . Actions create new forms, as well as "renewing" old forms by giving them new functions.

The inseparability of the system of objects from the system of axes is shown to us in the words of Milton Santos when he writes that: "On the one hand, the system of objects conditions the way in which actions take place and, on the other hand, the system of actions leads to the creation of new objects or takes place on pre-existing objects. This is how space finds its dynamics and transforms itself."[54] .

As for the historical conditioning factors, we can consider them to be the social and economic structure from which objects (form) and actions (function) emerge, and we can consider the evolution of this socio-economic structure (process) to be responsible for changing the morphology and meaning, obtained through interaction with society, of objects and actions[55] .

In this way, the historical conditioning factors, the socio-economic and cultural characteristics of each period, will give phenomenal existence to the objects arranged over the space of the basin and the actions carried out by and through them. However, as Milton Santos points out, we mustn't forget the importance of the techniques that affect the shape and distribution of the objects and the effectiveness of the actions[56] .

The imbricate fabric woven by the relationship established between the system of objects, the system of actions and the historical conditioning factors subjects the eye to perceiving socio-economic-spatial dynamics as a complex process, complementary and antagonistic by its very nature, which in the current historical period is permanently unfinished.

Some actions materialize in space through public policies[57] . These policies aim to provide the space with the infrastructure to make it viable or attract new investments, to establish certain networks (roads, sanitation, water distribution), to regulate and organize the space. In another sense, public policies are also aimed at reducing conflicts (social, of use) and socio-spatial differences, although in some cases they accentuate them.

The type of occupation of the area will be the result of a process that involves the interrelationships developed over time, with the consequent changes in historical conditions, between the distribution of objects and actions and with public policies aimed at each locality. This interaction results in the differential valuation of space, which translates into the differential rent of urban land[58] and land prices.

Differential income and the differential value of urban land will be one of the factors responsible for the distribution of social classes in the space. The wealthier classes will be located in areas with infrastructure and public facilities and the poorer classes in areas without them. Thus the

---

[52] SANTOS, Milton. Op. cit.

[53] "Action does not occur without an object; and when it is exercised, it ends up redefining itself as action and redefining the object" (SANTOS: 2002, 95).

[54] SANTOS, Milton. Op. cit. p. 63.

[55] "The evolution that marks the stages of the work process and social relations also marks the changes that have taken place in geographical space, both morphologically and from the point of view of functions and processes. This is how epochs are distinguished from one another" (SANTOS: 2002, 96).

[56] "Each and every historical period affirms itself with a corresponding list of techniques that characterize it and with a corresponding family of objects. Over time, a new system of objects responds to the emergence of each new system of techniques. In each period, there is also a new arrangement of objects. In reality, there are not only new objects, but also new forms of action" (SANTOS: 2002, 96).

[57] A set of positions taken by the state (public sphere of organized society) on problematic issues (OSZLACK, Oscar) and whose implementation processes generally take place via the state bureaucracy. It implies participation, plurality and freedom of opinion (GOMES, Edvania Torres Aguiar: 2004. Notas de aula).

[58] In urban areas, the differential income is obtained from the location, the existence of public facilities and the allocation of a certain amount of urban infrastructure. Another factor, but no less important, is the subjective valuation of space (RIBEIRO, 1997).

type of occupation, the social content of the different localities, implies public policies (residents' power of demand) and the quality of sanitation networks and garbage collection; it also influences the levels of sealing and deforestation, greater or lesser depending on the density of urbanization and the value placed by the community on the elements of nature.

Soil sealing can be considered a direct consequence of the production and reproduction of urban space and deforestation[59] . And, as described above, it will lead to an increase in the intensity of surface runoff, as well as an increase in the maximum flow and the anticipation of flood peaks; because it prevents the infiltration of water and decreases the time it takes to reach the watercourses, feeding them quickly[60] . The increased speed of water flowing over paved surfaces increases its erosive power.

Finally, the drainage network receives all the impacts resulting from the dynamics developed within the basin, regardless of which of the groups they develop in (the watershed group or the anthroposociological group), establishing the link between one and the other.

All the outputs of the watershed system will be concentrated in the drainage network, and its proper functioning is linked to the attributes of its inputs, which correspond specifically to the outputs of the other interrelationships. That said, we can conclude that the drainage network and its organization are one of the most sensitive elements of this system, receiving the result of all the variations in inputs and processes and being continuously altered by them. All the impacts linked to floods and their consequent economic and social losses come from it.

With the above, we do not intend to exhaust or represent all the relationships developed in the Tejipió river basin from our systemic representation, because as MONTEIRO (2000) warned "... the web of relationships between the elements and attributes of a part of the whole considered is often so complex that it is not possible to figure some of these 'combinations' in fact in a given landscape (or in a given geosystem)"[61] . And COELHO (2001) asserts the need to simplify the complex as something necessary for its understanding[62] .

Starting from an understanding of the Tejipió basin as a whole, we see the study of its parts and the relationships they maintain with each other as the only way of understanding this totality and analyzing its evolution. We seek to explain socio-environmental phenomena at the level of totality, opposing "the reductionist paradigm" which would look for it in the different localities dissociated from the environment, from the context in which they are inserted[63] and the holistic paradigm which takes a simplified view of the whole.

Despite having been developed within a positivist framework, the systemic method has dialectical elements at its core. MORIN (1999) points out, within the notion of a system, that all the relationships between the constituent parts of a system will be mediated by interactions. This gives us a dialectical view of the actions developed between complex units "made up... of interactions"[64] which, through the *feed-back* mechanism, will end up modifying each other. He also draws attention to "the need to think together, in their complementarity, competition and antagonism, the notions of order and disorder" which "raise exactly the question of thinking about the complexity of physical, biological and human reality."[65]

MORIN (1999) gives various definitions of order, sometimes referring to it as constancy, regularity, stability and repetition, sometimes as coercion, determination or causality that make phenomena obey the laws that govern them. At certain points, he defines it as logical coherence, which would open up the possibility of induction and deduction, and therefore of prediction; thus

---

[59] With the removal of the vegetation cover, due to the characteristics of the soil, the process known as "sealing" can begin, which consists of the clogging of pores, reducing permeability and therefore making it difficult for water to infiltrate the soil.

[60] BOTELHO, Rosangela Garrido Machado & SILVA, Antonio Soares da. Op. cit. p. 173.

[61] MONTEIRO, Carlos Augusto de Figueiredo. Op. cit. p. 37.

[62] Simplification consists of "... selecting what is supposedly most significant, avoiding what is uncertain and ambiguous..." (COELHO, 2001:37).

[63] MORIN, Edgar. Op. cit. 257.

[64] MORIN, Edgar. Op. cit. p. 264.

[65] MORIN, Edgar. Op. cit. p. 197.

revealing to us the "... universe assimilable by the mind"[66] , which would find in order the foundations of its logical truths.

Within a systems approach, order takes on a different meaning. Order becomes a necessary element for organizing the system and maintaining it, in which case "... the singular order of a system can be conceived as the structure that organizes it"[67] . We are working with this approach.

Like the notion of order, disorder will have different meanings in Edgar Morin's thinking. However, we see disorder as the ruptures that will force the structuring and restructuring of social classes and the economy, together with the ruptures that occur in physical dynamics as a result of human interventions.

Finally, we consider ruptures as destabilizations in the order or structure whose effects will be reflected in the actions and interactions carried out in the lakes drained by the Tejipió River and its tributaries, thus seeking to have a vision of the production and reproduction of the lakes as a dynamic process and, on the other hand, of the impacts caused to the physical substrate by the modifications implemented in it by the socio-spatial practices developed in it.

Subsequent chapters will analyze the physical and anthropic processes involved in the formation of the Tejipió river basin as a whole; although, throughout the dissertation, we repeatedly opt for the analysis of the parts, rather than the whole, as a method of apprehending reality, which would be undermined if we set out to do it from the whole, considering the dimensions of the Tejipió basin.

---

[66] MORIN, Edgar. Op. cit. p. 208.
[67] MORIN, Edgar. Op. cit. p. 198.

CHAPTER 2

Historical, geological and geomorphological milestones to help understand the process of formation and occupation of the Tejipió River Basin.

2.1 Evolution of the Recife Fluvial-Marine Plain: the process of formation of the Tejipió River Basin.

In Recife, what is not water, has been water or is reminiscent of water... The 'tyrant of water' has subjugated the land - water from the sea that covered it in a very remote era, water from the rivers that cut and cut it... water from the floods... water from the marshes that the vegetation of the mangroves shadows and hides, water from the sea that doesn't capitulate before the reefs and returns, twice a day, to visit its lost domains through the arms of the rivers. "

Waldemar de Oliveira

In order to understand the formation of the Tejipió river basin, it is necessary to understand the process that gave rise to the Recife fluvial-marine plain, since the aforementioned basin is largely superimposed on this geomorphological unit.

Previous studies on this subject (OLIVEIRA, 1942; SUGUIO et. al. 1985; DOMINGUEZ et. al., 1990; LIMA FILHO et. al., 1991; COUTINHO et. al., 1998; ALHEIROS, 1998; among others) attribute the origin of the Recife plateau to the variations in sea level that occurred during the Quaternary Period as a result of global climate change. During this period, there were alternating glacial and interglacial periods, in the former of which the sea level dropped due to the containment of water in the polar ice caps.

Thus, each period of warming or melting corresponds to a phase of marine transgression which, in turn, slows down the surface flows that reach the flooded areas and, therefore, sedimentation. In the case of the eastern Brazilian coast, Quaternary transgressions carved out cliffs and gave rise to lagoons at the foot of these elevations. As a result, during glacial periods, due to the return of the sea, local base levels were reduced and erosion resumed. In some of these periods, during the Pleistocene and Holocene, sediments were deposited in coastal areas, building up "by the aggregation of successive sandy coastal strands"[68] the marine terrages (or beaches).

In addition to the deposits from oceanic processes, the Recife plain was also formed by fluvial sedimentation from the penultimate transgression onwards, changing the characteristics of the material that makes it up as it was reworked by the rivers.

However, LIMA FILHO et. al. (1991) draws attention to the fact that this plateau is "undoubtedly of tectonic origin"[69] , in its early days having close links with the Pernambuco Lineament[70] and with the opening of the Atlantic Ocean. According to the aforementioned author, the rotational movement of the Northeast microplate would have pushed this fault to the north, making it possible, from this movement, to form a depression that would later form Recife Bay. However, this depression would only have been limited to the north of the lineament; the entire southern part of the lineament, belonging to the Cabo sub-basin, would only have been an extension of the bay.

---

[68] COUTINHO, Roberto Quental et al. *Climatic, Geological, Geomorphological and Geotechnical Characteristics of the Dois Irmao Ecological Reserve.* In: MACHADO, Isabel Cristina et. al. *Reserva Ecológica de Dois Irmaos: Estudos em um Remanescente de Mata Atlántica em Área Urbana (Recife - Pernambuco - Brazil).* Recife: Editora Universitária da UFPE, 1998. p.35.

[69] LIMA FILHO, Mario Ferreira de et. al. *Origin of the Recife Plain.* In: Estudos Geológicos UFPE/DEGEO (Series B - Estudos e pesquisas). Revisao Geológica da Faixa Costeira de Pernambuco, Paraíba e Rio Grande do Norte. UFPE: Recife, 1991. vol. 10, p. 157-176.

[70] An E-W shear zone formed around 600 million years ago (Precambrian period, Brasilian age) as a result of tangential stresses on rocks of the crystalline basement. It forms a strip, a few kilometers wide, of differentiated structure and resistance that stretches across the entire state of Pernambuco and continues into the African continent. It has generated an important structure whose effects have been felt in the development of the relief, the Serra das Russas being a case in point (adapted from ALHEIROS: 1998, 51).

The combination of tectonic and sedimentary factors leads LIMA FILHO et. al. (1991 & 1998) to consider the Recife plain a complex phenomenon. And precisely because of its complex formation, the current plain fits in better with the most recent geological history, the Quaternary[71] . The geological history of the Recife plain will be told here from the deposition of the Barreiras Formation, based on the studies developed by SUGUIO et. al. (1985), LIMA FILHO et. al. (1991) and COUTINHO et. al. (1998); since events prior to this event are difficult to reconstruct.

In order to better visualize the process of formation of the Recife plain, it was divided into eight phases, according to the major events during the Quaternary period, with the exception of the deposition of the Barreiras Formation at the end of the Tertiary period.

> **Phase 1 - Sedimentation of the Barreiras Formation:** At the end of the Pliocene, under semi-arid climate conditions, subject to concentrated and torrential rains, the Barreiras Formation was sedimented over an extensive stretch of the Brazilian coast, as well as over the current submerged continental shelf, due to the sea level being lower than it is today[72] . This lithostratigraphic unit was deposited in the form of proximal alluvial fans and fluvial deposits of interlocking channels, with fluvial-laguar facies and alluvial plains[73] (figure 5a).

> **Phase 2 - Maximum Transgression:** After the sedimentation of the Barreiras Formation, the climate became wetter and the sea level began to rise (the maximum transgression), which eroded the outer portion of this formation, carving out cliffs[74] (figure 5b).

> **Phase 3 - Post-Barrier sedimentation:** Consequence of regression. In this phase there is the deposition of continental sediments, in the form of alluvial fans, at the foot of the cliffs through the channels excavated in the Barreiras deposits. However, these fans are not visible on the Recife plain[75] (figure 5c).

> **Phase 4 - Penultimate Transgression:** During this phase the deposits from the previous phase were reworked and covered[76] , with the waves reaching the cliffs from the maximum transgression. With the rise in sea level, the lower river courses were drowned and transformed into estuaries and lagoons[77] (figure 5d).

> **Phase 5 - Construction of Pleistocene Marine Terrages:** with the regression resulting from the transgression of the previous phase, Pleistocene marine terrages were built, formed from prograding beach ridges. In this phase, the terrages filled the entire bay of Recife and Jaboatao dos Guararapes[78] (figure 5e).

Possibly, during this phase, there was a change in the course of the Capibaribe River in the vicinity of the Sao Joao da Várzea waterhole, moving, according to LIMA FILHO et. al. (1991), from the direction S 30° E to N 40° E, always obeying old fault lines. This change in the course of the Capibaribe has led to the erosion of part of the Pleistocene marine terrace, with the consequent installation of marshes and/or lagoons on the plain, as well as the construction of fluvial terraces[79] .

Based on this, we have put forward two hypotheses to be developed in further work, as they do not fit the objectives of the work in question.

The first deals with the possibility that the Capibaribe River, in periods before the Pleistocene regression, flowed in the current bed of the Tejipió River or in its vicinity and, with the change of

---

[71] According to DOMINGUEZ et. al. (1990); LIMA FILHO et. al. (1991).

[72] SUGUIO, Kenitiro et al. *Relative sea level fluctuations during the upper Quaternary along the Brazilian coast and their implications for coastal sedimentation.* Revista Brasileira de Geociencias, vol. 15, n° 4, p.273-286.

[73] COUTINHO, Roberto Quental et. al. Op. cit. p.35.

[74] SUGUIO, Kenitiro et. al. Op. cit. p.282.

[75] COUTINHO, Roberto Quental et. al. Op. cit. p.35.

[76] COUTINHO, Roberto Quental et. al. Op. cit. p.35.

[77] SUGUIO, Kenitiro et. al. Op. cit. p.282.

[78] COUTINHO, Roberto Quental et. al. Op. cit. p.35.

[79] LIMA FILHO et. al. Op. cit. p.173.

course, the Tejipió began to take over the drainage of the area that today corresponds to its basin[80] .

The second hypothesis points to the possibility of the Tejipió in the past having been a tributary of the Capibaribe, which, with an uplift possibly of tectonic origin at the height of the Sao Joao da Várzea squat, became separate from the Capibaribe heading south[81] .

The uplift mentioned above can also be considered as a possible cause of the change in the course of the Capibaribe, being responsible for the 60m rise in the small stretch that separates the course of the Capibaribe River from that of the Tejipió. Map 1 shows the hypsometric differences in the Tejipió basin.

> **Phase 6 - Last Transgression:** In this phase, the drainage network is drowned by the rise in sea level. Barrier islands were then formed which, added to the regressive sand ridges formed in the previous phase, isolated the lagoons. Inside the latter, a process of sand and clay deposition began, partly due to the reworking of Pleistocene sands and partly due to the action of the rivers (formation of intralaguar deltas)[82] (figure 5f).

> **Phase 7 - Construction of Intralaguar Deltas:** In the lagoons located at the mouths of the main rivers that flowed into the Atlantic Ocean, intralaguar deltas were formed, fed essentially by fluvial sediments[83] (figure 5g).

> **Phase 8 - Construction of the Holocene Marine Terrace:** The formation of the Holocene marine terrace takes place as a result of the lowering of the relative sea level after the transgressive maximum of 5,100 B.P. This marine terrace originates from the deposition of sediments on the barrier islands, resulting in the progradation of the coastline[84] (figure 5h). The ancient barrier islands are also responsible for the formation of the sandstone reefs[85]

This process of formation, through which the Recife plain passed, left the Tejipió River basin with the geological units that make it up (see map 2). Within the domain of this basin we can find the following lithostratigraphic units: crystalline basement, Cabo Formation belonging to the Pernambuco Basin, Barreiras Formation and the "Quaternary formations" which encompass "all geological units formed in the Quaternary Period, regardless of their nature and genesis"[86] .

In the following pages we will discuss the main characteristics of each of the geological records found in the basin in question.

---

[80] According to an informal discussion with LIMA FILHO, 2005.
[81] According to an informal discussion with CORREA, 2004.
[82] LIMA FILHO et. al. Op. cit. p.173.
[83] SUGUIO, Kenitiro et. al. Op. cit. p.283.
[84] SUGUIO, Kenitiro et. al. Op. cit. p.283.
[85] COUTINHO, Roberto Quental et. al. Op. cit. p.36.
[86] ALHEIROS, Margareth Mascarenhas. *Landslide Risks in the Metropolitan Region of Recife*. UFBA: Postgraduate Course in Geology. Doctoral Thesis, 1998, 135pp.

MAP 1: HYPSOMETRIC MAP OF THE TEJIPIÓ RIVER BASIN

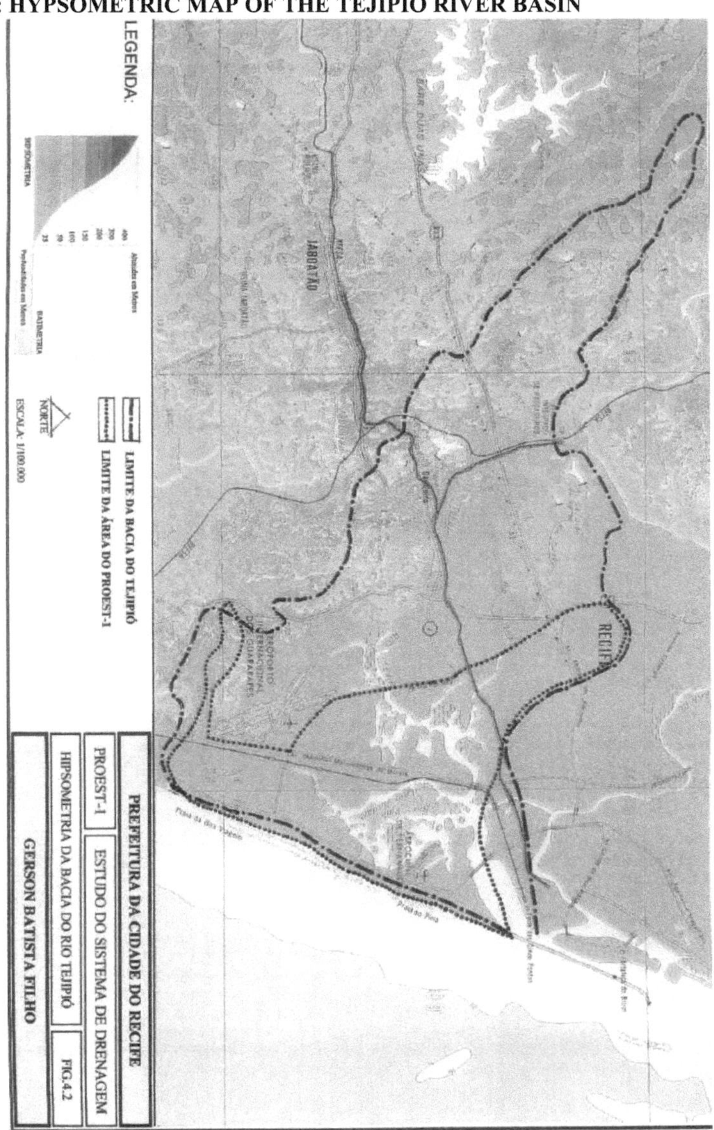

Source: PCR/SEPLAN/PROEST-1/1996; FIDEM/1986.

FIGURE 5: PALEOGEOGRAPHIC EVOLUTION OF THE REEF PLAIN

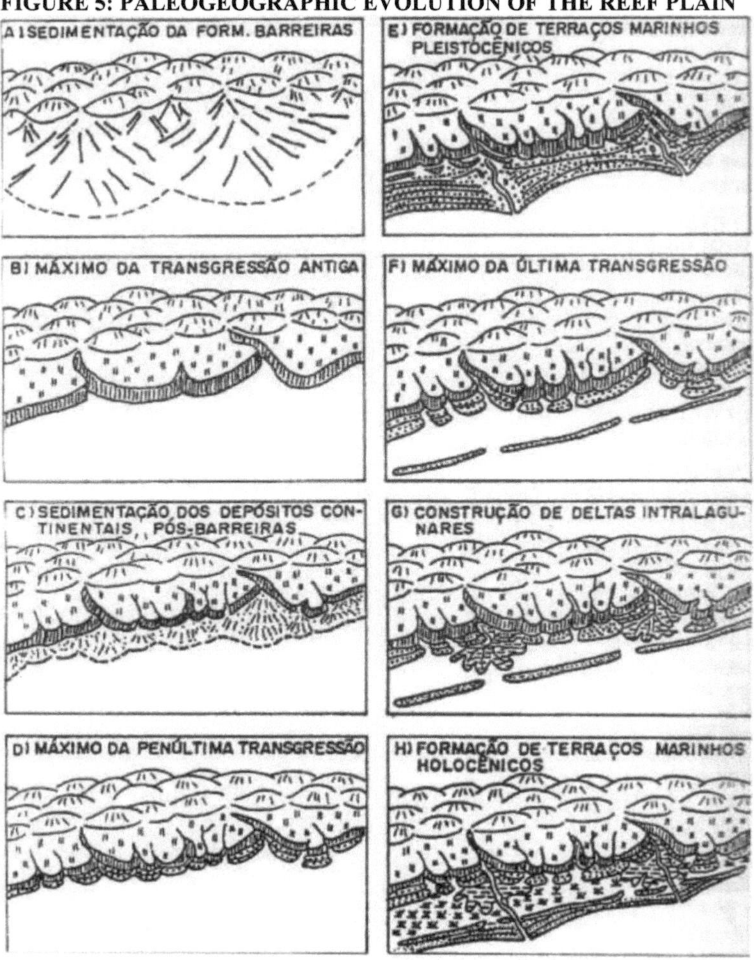

Source: SUGUIO et. al. 1985.

The Crystalline Embayment is the oldest of the geological records located in the basin. It represents the crystalline rock substrate of the Gneissic-Migmatitic Complex, of Archean age (2.1 to 1.5 billion years old), belonging to the Pernambuco-Alagoas Massif[87] . It is sotoposed throughout the basin and is made up of Precambrian granitic rocks, around 2 billion years old. These rocks outcrop in some stretches located to the west of the basin, in its upper course; as well as, in some stretches, in the riverbed itself (see photo 1)[88] . Under local climatic conditions, the residual soil resulting from the weathering of this material is clayey, although it contains a large quantity of dark minerals (ferromagnesians), which, together with feldspars, have a strong tendency to clayize[89]

Photo 1

Outcrop of igneous rock in the Tejipió riverbed. Author: Paulo Tavares, 2005.
[87] BRITO NEVES, Benjamin Bley de. *The Geological Map of the Eastern Northeast of Brazil, scale 1:100,000.* USP: Institute of Geosciences. Thesis, 1983, 177pp.
[88] The photo also shows the intense occupation of the banks of the Tejipió river in the stretch where it serves as the boundary between Recife and Jaboatao dos Guararapes. The left bank, Recife, gives the impression that the river has been canalized, even though the Tejipió has not undergone any structural intervention.
[89] ALHEIROS, Margareth Mascarenhas et al. *Geological Map of the City of Recife.* SEPLAN/Prefeitura da Cidade do Recife. 1995.

To the south-east of the Cristalino, the Cretaceous Sedimentary Basins were laid down on blocks of the embayment lowered by geological faults that occurred during the separation of the South American and African tectonic plates approximately 100 million years ago (Cretaceous period, Aptian age)[90]. In the Pernambuco Basin, conglomerates, feldspathic sandstones (arcsandstones) and claystones were deposited, forming the so-called Cabo Formation. Because they originate from fragments of the crystalline basement, these sedimentary rocks also developed residual soils that are very clayey (Podzolic)[91], but the areas where they predominate have a lower population density, and their dismantling is mainly linked to the extraction of material for civil construction (a process that can be seen on the banks of the BR 101 highway, in Barro).

Indistinctly covering the crystalline basement and the units of the Cretaceous Sedimentary Basins are the sediments of the Barreiras Formation, deposited 2 million years ago during the Tertiary period (Pliocene age)[92], with "their deposition associated with Cenozoic events of a climatic and/or tectonic nature"[93]. The Barreiras Formation is made up of sandy deposits of fluvial origin, deposited under high-energy conditions, sometimes covered by layers of sandy and clayey sediments, generated by successive floods (debris flows). Barrier sediments are more common outside our study area, in the hills of northwest Recife.

The fluvial-marine plain, in turn, is formed by three geological records, according to ALHEIROS et. al. (1995): the Pleistocene Marine Terrace, the Modified Pleistocene Marine Terrace and the Holocene Marine Terrace. These geomorphological units are the result of the deposition of material by the two main modeling agents of this relief, the rivers and the ocean, during the processes of marine transgression and regression that occurred in the Quaternary period and correspond,

---

[90] ALHEIROS, Margareth Mascarenhas et al. Op. cit.
[91] ALHEIROS, Margareth Mascarenhas et al. Op. cit.
[92] ALHEIROS, Margareth Mascarenhas et al. Op. cit.
[93] ALHEIROS, Margareth Mascarenhas. Op. cit. p.60.

together with the alluvium and the lagoon, deltaic and estuarine sediments, to the "Quaternary formations".

During the Pleistocene epoch, the Pleistocene Marine Terrace was formed, a beach around 100,000 years old. This is a flattened geomorphological unit which, in contrast to the hills that surround it, has low altimetry levels of between 7m and 10m. It is made up of light quartz sands, which are unconsolidated on the surface, but become more compact and darker in depth, as a result of the Podzol formation process which leads to the cementation of the sand by humic acid and iron oxide, giving cohesion to the sandy material[94] (see photo 2).

Photo 2

Outcrop of Pleistocene marine terrace in the Areias neighborhood. In the background is the Ignes Andreazza Residential Complex. Author: Paulo Tavares, 2005.

The Modified Pleistocene Marine Terrace, on the other hand, corresponds to an ancient Pleistocene beach that has been profoundly altered by the rivers. Thus, its behavior in the subsurface is very irregular, sometimes dominated by reworked sands, sometimes by deposits of soft, organic clay, deposited on the old fluvial flood plains.

The Holocene Marine Terrace represents the current strip of beach, located between the Setúbal channel and the coastline, which was established approximately 5,000 years ago, during the Quaternary, Holocene epoch. This well-individualized geomorphological unit has elevations of between 3 and 5 meters and is distinguished from the Pleistocene Terrace by the absence of dark cement at the base and the presence of shell fragments[95] . Today, from the point of view of urban land valuation, it is one of the most expensive square meters in the city, with various agents working to promote land valuation and increase profits from the commercialization of urban land. On the other hand, the drilling of wells without a pre-defined distance between them, and the failure to rationalize the use of underground water near the salt wedge, has led to the contamination of aquifers by salt water.

In the low-lying areas corresponding to the estuary of the basin's component rivers, we find alluvium and ancient and recent lagoonal, deltaic and estuarine sediments, which are grouped together under the name of Fluvial-Laguar Deposits.

The alluvial fans are dominantly sandy and are located along the channels. They have clayey sediments with the presence of organic matter and are deposited on the flood plains during channel overflows. The lagoon, deltaic and estuarine sediments have a varied sandy-silty-clay composition

---

[94] ALHEIROS, Margareth Mascarenhas et al. Op. cit.
[95] ALHEIROS, Margareth Mascarenhas et al. Op. cit.

with organic matter, and show plane-parallel stratification[96] .

Due to sea level fluctuations, these deposits can intermingle typical lagoon sediments, rich in shells, with freshwater sediments deposited in lagoons, marshes and swamps. Locally, there are subsurface layers of soft clay, diatomite and peat, respectively from these environments[97] .

2.2 Destabilization and Restabilization of the System: the process of human occupation.

"Everything flows, nothing persists or remains the same."

Heraclitus of Hephaesus

Changes in technology and the production system have, over time, led to metamorphoses in the territorial configuration of the city of Recife. These changes, together with other factors, have influenced the distribution of objects and actions in the different regions of the city. Among these regions is the Tejipió river basin, the subject of this study, which contains, within its boundaries, different socio-cultural and economic realities, as well as different forms of urban environmental impacts, whose origins are linked to the historical process developed under the aegis of the capitalist mode of production[98] .

The forms of appropriation of the land in this basin have varied in different historical periods, acquiring different characteristics in each one.

Occupation began in the 16th century with a small settlement near the mouth of the main river, where the Afogados neighborhood is now located, at the same time as the urban site of Recife.

In the middle of the century, sugar mills were set up on the alluvial floodplains of the Tejipió and Jiquiá rivers. During this period, both rivers played an important role in transporting the sugar produced to the port of Recife.

However, from the sugar mills emerged the peripheral population nodes which, with the progression of urbanization in the 19th century, would become the city's neighbourhoods and suburbs. Land development, an integral part of the process of transforming rural areas into suburban and urban areas, would subdivide the large sugar cane estates into farms (which combined residential use with fruit plantations) and, at a later stage, the farms were divided into residential plots through subdivisions carried out by the owners and by people or companies who bought them[99] .

The accelerated population growth seen at the beginning of the 20th century, when migration to Recife intensified, caused the city's urban population to almost quintuple in the first fifty years[100] , and the consequent urban expansion led to the occupation of the area drained by the Tejipió, through spontaneous settlements or allotments promoted by private owners or through government intervention (IPSEP, Vila das Lavadeiras, Vila da SUDENE, among others).

That said, by way of an introduction, we will look at the history of the occupation of the Tejipió river basin, in an attempt to provide a better basis for understanding the current situation in the basin.

2.2.1 Occupation of the Tejipió River Basin

During the 16th century, the Tejipió sugar mill was set up on the right bank of the main river in the basin, later converted into a cattle ranch, and on the banks of the Jiquiá the Jiquiá, Curado and Sao Paulo sugar mills. The demarcation of the lands of the then Sao Timóteo mill, later called Santo Antonio do Jiquiá, dates back to 1598, "in satisfaction of the respective letter of sesmaria granted by the captaincy's grantees"[101] , although it had already existed previously. The lands of the Jiquiá mill occupied an area that today corresponds to the neighborhoods of Jiquiá, Areias, Cagote, Estancia and

---

[96] ALHEIROS, Margareth Mascarenhas et al. Op. cit.

[97] ALHEIROS, Margareth Mascarenhas et al. Op. cit.

[98] Despite the fact that the colonization system established in Recife was based on large estates, monoculture and slave labour, it excluded the determinants of capitalist production, such as the primary and antagonistic relationship between capital and labour (BARROS FILHO: 2000 38-39).

[99] MELO, *MárioLacerda de. Meiropolizacao e Subdesenvolvimento: o caso do Recife*. Recife: UFPE, 1978. p.64.

[100] MELO, Mário Lacerda de. Op. cit. p.70-71.

[101] PERREIRA DA COSTA, Francisco Augusto. *Around Recife*. Recife: Fundaçâo de Cultura Cidade do Recife, 1981. p. 85.

Ipiranga.

The sugar mills integrated this area into the production system of the time by implementing the predominant use of arable land in the Captaincy of Pernambuco. During this period, the rivers mentioned, which were then navigable, were used to transport sugar to the port of Recife. The sugar was bagged and deposited in barges from the Jiquiá and Afogados passes.

The Passo de Santa Cruz do Jiquiá, a small trading post, was founded after the Restoration of Pernambuco (17th century); it consisted of a wharf located at the mouth of the river of the same name (at the confluence point of the Jiquiá River with the Tejipió) "where transport vessels freely arrived"[102] .

In addition to sugar, this warehouse also shipped wood and other marketable products to the so-called Recife plague, as well as receiving and storing products that were destined for the various mills (Tejipió, Peres, Sao Paulo, Curado) and towns in their immediate vicinity.

Thus, with the advent of occupation and economic exploitation of the land in the basin at the end of the 16th century, environmental conditions began to change with the spread of sugar cane over the alluvial floodplains of the Tejipió and Jiquiá; with the concomitant deforestation of the humid forest, which predominated over these floodplains and the hills, also cleared for the benefit of this crop. In addition, the deforestation of the mangroves that extended over large areas of the plain, which would gradually be landfilled, began.

On the other hand, the primary use of the floodplains of the aforementioned rivers can be considered to have had little impact compared to the other uses that have succeeded it throughout history. The implementation of sugarcane has resulted in the reduction of native vegetation cover and the acceleration of some erosion processes followed by the silting up of liquid bodies, but within acceptable levels for the local water system, i.e. below the degree of resilience[103] which guarantees that the system will not be destabilized.

In this initial period of occupation, social practices were developed taking into account the characteristics of the physical environment, the elements of nature were part of the daily lives of individuals; with the technical and technological increase, practices become increasingly independent of the environment and society turns its back on nature and its elements. As time goes by, the pace of life accelerates, transforming nature into a scenic backdrop on which objects are arranged; turning it into an inanimate stage for human actions, which start to take place with their backs to the environment in which they take place.

While in the areas close to the center urban occupation raged throughout the 16th, 17th and 18th centuries, receiving a major boost from the Dutch invasion of 1630, it was incipient and sparse in the regions located to the south and southwest of the port.

However, it is incorrect to think that the process of occupation took place solely from the port inland, since, in parallel with the development of the areas close to the port, there was also a process of occupation in the opposite direction, resulting from the significant transformations taking place around the city[104] . At the time, the mills were "autonomous centers of life"[105] , some of which became large population centers, with around 100 to 200 residents.

The crisis of sugar production in the 18th century was one of the factors responsible for the subdivision of the large rural properties located on the outskirts of Recife. Initially, these were divided into sites which, from 1840 onwards, were subdivided into smaller plots.

---

[102] PERREIRA DA COSTA, Francisco Augusto. Op. cit. p. 88.

[103] Resilience is considered to be the system's capacity to absorb impacts, i.e. the system's ability to absorb and adapt to changes in its order (structure) without these destabilizing it. When the modification (disorder) is lower than its degree of resilience, the system absorbs it without major consequences for its stability or functioning; when it is higher, it causes ruptures leading to restructuring or complete destabilization of the system (CORREA, Antonio Carlos de Barros: 2004. Notas de Aula).

[104] "The view of the settlement of Recife is not correct if we see it only as a movement from the port area to Antonio Vaz Island. There was an equally important - economically important - movement from the interior towards the port. I'm referring to the movement that originated in the sugar mills in the mid-16th century..." (MELLO apud BARROS FILHO: 2000, 37).

[105] CASTRO, Josué de. *Ensaios* de *Geografia Humana*. Sao Paulo: Brasiliense, 1966.

This situation of incipient and dispersed occupation of the Tejipió basin began to change in the first decades of the 19th century. With the elevation of Recife to the capital of the Province of Pernambuco in 1823, a process of urban restructuring began with the opening up of streets, the construction of new bridges and carriage roads that connected the interior to the central core, works that enabled urban expansion to the southwest: Afogados, Jiquiá and Tejipió. In addition, railroads were installed to run streetcar lines and trains to more distant locations. In order to consolidate these constructions, several landfills were made on the banks of rivers, in mangroves and wetlands, making this a common urbanization practice in Recife[106] .

New settlements sprang up along the roads and railroads that led from the central area towards the interior, such as Barro (formerly called Barro Vermelho), along Estrada da Vitória (now Av. Dr. José Rufino), and Nossa Senhora da Boa Viagem, served by a railway station of the Estrada de Ferro Sao Francisco[107] . In other settlements, the existence of which dates back to periods before those mentioned, occupation intensified after the construction of these roads, as in Tejipió and Jiquiá. These settlements, located on the outskirts of the city of Recife, were called nodules or peripheral settlements by Mario Lacerda de Melo[108] .

We can consider the existence of the nodules of peripheral occupation, with growth taking place along the roads, a sketch of the tentacular expansion that would take place in the city, assuming intra-urban dimensions[109] , at first, and inter-urban, during the process of metropolization.

The construction of the Estrada da Vitória, also known as the general road, stands out in the urbanization process of Recife's southwestern region. By facilitating communication with the central core, it led to the emergence and consolidation of a rosary of settlements along its banks. The subsequent densification of these settlements gave rise to the various neighborhoods that make up this region.

The construction of the general road linking the center of Recife to Santo Antao - today Vitória de Santo Antao - was carried out by order of governor Luís do Rego Barreto in 1819, with the Jiquiá embankment or road as its starting point[110] . The installation of this object changed the dynamics of the region, making the village of Jiquiá, in 1831, a "frequented point of passage"[111] with accommodation for the train drivers and their pack animals. In the vicinity of this settlement there were several farms on land belonging to the engenho.

During this period, the Santa Cruz do Jiquiá pass was a small but busy village, due to the activities that took place there (trading and storage). These activities increased considerably with the construction of the first section of the Vitória road, completed in 1836, linking Povoaçao dos Afogados to Areias.

With the completion of the first section of the road, the village of Barro Vermelho emerged, near Engenho Peres, and the village of Tejipió gained a new population, given the ease of communication with Recife, from the occupation of the lands of the former Engenho Peres[112] .

The urban development of the localities that make up the basin takes place slowly, with the neighbourhoods organizing themselves according to the regularity of their radii of coverage[113] and,

---

[106] Most of the landfills were carried out slowly and gradually by both public authorities and private individuals, providing another feature in the expansion of Recife's sprawl: if the city was growing from the port to the interior, it was now moving in reverse.

[107] One of the first railroads in the country connected Recife to Cabo de Santo Agostinho. Its name comes from the fact that in the original project it was supposed to extend as far as the River San Francisco. One of its stations, the one located in Boa Viagem, became the "gateway" to this locality with the beginning of its demand for summer vacation, "according to the 'salty baths' fashion, which would come to replace the old 'river baths'" (GOMES: 1997, 220).

[108] MELO, Mário Lacerda de. Op cit. p.62.

[109] "In the mid-19th century, Recife clearly showed a radial configuration, made up of a central nucleus close to the port and a series of small villages on the banks of the rivers, along the main transportation lines that converged on the port." (BARROS FILHO: 2000,38).

[110] PERREIRA DA COSTA, Francisco Augusto. Op. cit. p. 89.

[111] PERREIRA DA COSTA, Francisco Augusto. Op. cit. p. 89.

[112] RECIFE, City Hall. *Municipal Profile-History and Urban Evolution*. Recife: Directorate of Urban Planning, 1989. p. 159-160.

[113] RECIFE, City Hall. Op. cit. p. 159-160.

in this way, Tejipió, Coqueiral, Barro, Pacheco, Sancho, Totó, among others, "gradually become denser and gradually take on the upward physiognomy of the city."[114]

The occupation of these peripheral nodules only changed significantly in the first half of the 20th century, when the flow of migrants to Recife increased (between 1890 and 1910, the population grew by 39%). This led to their spatial growth, making them increasingly compact. A case in point is Boa Viagem, which emerged as "a new densely occupied strip (...) along the south coast..."[115] due to the amenities of the coast.

The neighborhood of Boa Viagem has its origins in the settlements of fishermen from the village of Barreta at the end of the 17th century (a large part of the area occupied by Boa Viagem belonged to Barreta) and until the end of the 19th century the strip of beach remained occupied by fishermen and religious institutions[116].

For a long time Boa Viagem was considered just a salty bathing resort, with characteristics that set it apart from the others, such as a calm sea, a low, sandy beach and a gentle slope, making it the bathing resort most frequented by the people of Recife. The life of the town was therefore linked to the bathing area in the summer, while the rest of the year it was practically uninhabited.

Boa Viagem became part of Recife's urban development at a late stage, around the beginning of the 1920s, when bridges and new roads were built to make it easier to connect the southern stretch of beach with the rest of Recife. According to GOMES (1997), this fact "refers to the physical-natural limitations provided by the Pina basin, which prevented the connection [by] land of this coastal strip to the spaces of consolidated occupation in and from the primitive central nucleus of Recife"[117].

The maintenance of accelerated population growth in the second half of the 20th century has as one of its reflexes the constant increase in built space, incessantly incorporating new portions of the municipal territory with the urban "increasing its territorial area with the inclusion of areas with rural characteristics"[118], as well as green areas not yet occupied, through the expansion of deforestation and soil sealing during the incorporation process.

MELO (1978) draws attention to the fact that this population increase "... would entail a corresponding expansion of the urban space and produce substantial changes in the configuration of this space and in the urban structure itself in general"[119].

Thus, the city's growth began to intensify over the natural space of the waters, with the expansion of the occupation of the banks of the rivers and mangroves through landfills, initially carried out in the central districts, which then spread to other areas of the city, causing serious changes in the drainage network, such as the disappearance of some channels[120] and the artificialization of others.

This process of expanding occupation brings with it the overvaluation of the areas closest to the center and the areas originating from the subdivision of old mills, disputed by the layers of higher purchasing power, leaving the areas "devoid of economic value" relegated to occupation by less well-off people, resulting in the intensification of the occupation of swampy and flooded land, taken over by mangroves and rivers by a large contingent of low-income people to build houses on land that had been filled in, often with insufficient elevations; They also occupy the hills and the voids between the wide urbanized radial spaces[121].

This new contingent, faced with the need to locate in the urban space, is faced with "the problem of how and where to live"[122]; since the differential value of urban land becomes a limiting factor to their free movement in this space, given that private ownership of the land means that there

---

[114] RECIFE, City Hall. Op. cit. p. 159-160.
[115] MELO, Mário Lacerda de. Op. cit. p. 71.
[116] GOMES, Edvania Torres Aguiar. *Landscape Cut-outs in the City of Recife: a geographical approach*. Sao Paulo: USP/CDG. PhD Thesis. 1997. p. 220.
[117] GOMES, Edvania Torres Aguiar. Op. cit. p. 220.
[118] COSTA, Eda Maranhao Pessoa da. *Urban Expansion and Spatial Organization*. Recife: UFPE, 1982. p. 67.
[119] MELO, Mário Lacerda de. Op. cit. p.71.
[120] Here understood as river channels responsible for the flow of surface water.
[121] MELO, Mário Lacerda de. Op. cit. p.72.
[122] CORRÊA, Roberto Lobato. *Geographical Trajectories*. Rio de Janeiro: Bertrand Brasil, 1997. p.132.

is a land price to be paid and that this is regulated by the profitability of capital in current and future use[123] .

The demand for land was increased by the population growth caused by migration. State planning came into play from the 1960s onwards with large housing developments and the construction of the COHABs.

The population dynamics exposed in this sub-chapter, referring to those developed in the Tejipió river basin, are no different from those evidenced in other parts of the city, since the basin itself is just one part of the complexity of the whole represented by the city of Recife.

---

[123] RIBEIRO, Luiz Cesar de Queiroz. *From Tenements to Closed Condominiums: The Forms of Housing Production in the City of Rio de Janeiro.* Rio de Janeiro: Civilização Brasileira, 1997. p.77.

CHAPTER 3

Assessment of Complex Systems: the analysis of socio-spatial dynamics in interface with environmental dynamics in the Tejipió River Basin.

3.1 Tejipió River Basin: Considerations on the Characteristics of a Complex System.

"We can't enter the same river twice, because its waters are renewed every moment. We don't touch the same being twice, because it continually changes its condition."

Heraclitus of Hephaesus

Arising from the concern to analyze the evolution of occupation and the main impacts caused by the modifications implemented in the areas that make up the Tejipió river basin since the second half of the 20th century, this study aims to understand the reality constructed from the interaction between man and natural elements, based on a systematic view of these interactions and the resulting changes, analyzing the socio-environmental dimensions of the area, taking into account all the component factors of this complex system, made up of physical-natural, historical and socio-economic elements in constant interaction. This means that physical dynamics are not dissociated from socio-spatial dynamics and vice versa. It also seeks to emphasize the urbanized areas of the basin located within the city of Recife.

This first sub-chapter will attempt to describe the main characteristics of the Tejipió river basin. To do this, it is necessary to characterize the drainage network of the Pernambuco coast, corresponding to the Recife Metropolitan Region (RMR), and its division into hydrographic basins in order to understand the basin in question within its universe[124] .

The hydrographic network of the Recife Metropolitan Region is made up of ten river basins: Botafogo, Igarassu, Timbó, Paratibe, Beberibe, Capibaribe, Tejipió Jaboatao, Pirapama and Ipojuca (see map 3). In most of these basins, the main river has its sources in the hills of the Zona da Mata, still within the boundaries of the RMR or in nearby municipalities, with the exception of the Capibaribe and the Ipojuca, whose sources are located in the foothills of Borborema, in the Agreste Mesoregion. Therefore, these basins are made up of rivers that are not very long and are small in size, making it necessary to group them into coastal basin groups (GL) for planning purposes.

The Government of the State of Pernambuco uses the river basin as the geographical unit for planning, evaluating and controlling water resources. Based on the state basins, Planning Units (UP) were established[125] , 29 in all, made up of the main basins and, eventually, a group of two or more basins, formed due to the small size of each of these geographical units.

The Pernambuco coastline, due to the characteristics of its hydrography, is subdivided into six Groups of Small Coastal River Basins, two of which are located in the RMR: GL1 and GL2.

---

changes due to the behavior of that particular system". In this specific case, the drainage of the RMR coastline is subject to similar physical conditions, which means that the processes seen in the other basins are similar to those observed in the Tejipió basin; the latter is distinguished from the others by the attributes imposed on it by human activities.

[125] The Planning Units (PU) are made up of the 13 basins considered to be the most important in the state and 16 groups of basins, 6 Groups of Small Coastal River Basins (GL), 9 Groups of Small Inland River Basins (GI) and 1 group made up of the small rivers that drain the Fernando de Noronha archipelago.

MAP 3: MAP OF THE REGION'S RIVER BASINS
METROPOLITAN RECIFE

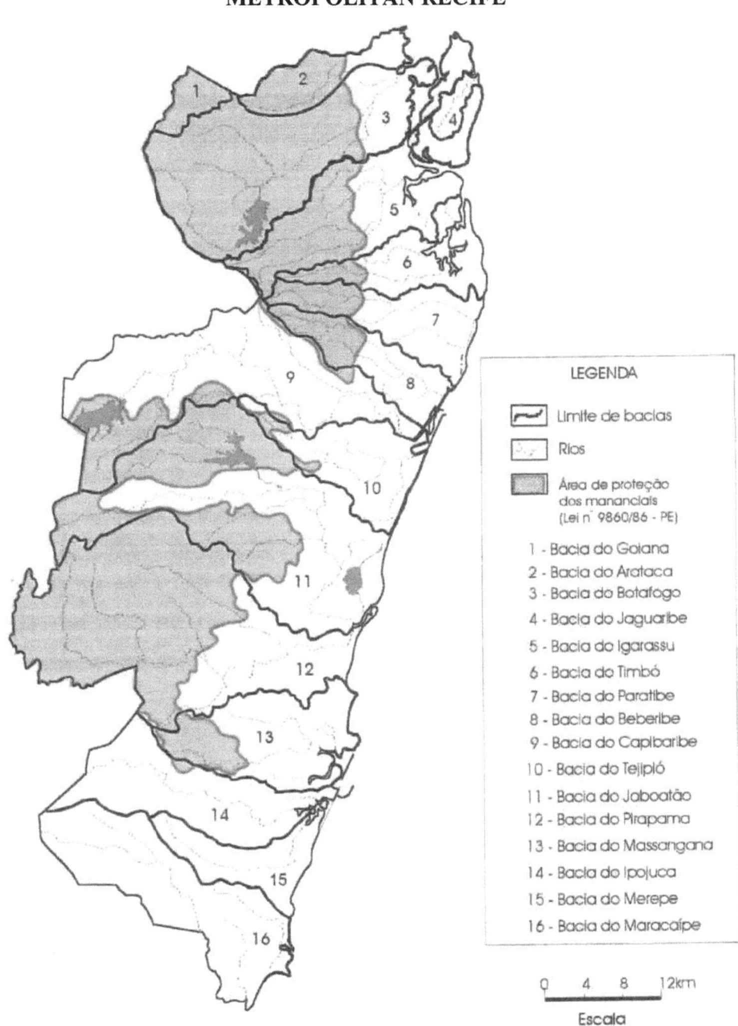

source: *FIDEM (1995)*
Adapted from ALHEIROS: 1998.

GL1, located in the north of the Metropolitan Region, is formed by the Botafogo, Igarassu, Timbó, Paratibe and Beberibe river basins. It is bordered to the north by the Goiana river basin, to the south by the Capibaribe basin, to the west by the Capibaribe and Goiana river basins and to the east by the Atlantic Ocean, draining areas in the municipalities of Itamaracá, Itapissuma, Aragoiaba, Igarassu, Abreu e Lima, Paulista, Camaragibe, Olinda and Recife (see map 4).

GL2 is located to the south of the Capibaribe basin, which is made up of the Jaboatao and Pirapama river basins. It is bordered to the north and west by the Capibaribe basin, to the south and west by the Ipojuca river basin and to the east by the Atlantic Ocean. It drains the municipalities of Jaboatao dos Guararapes, Moreno, Cabo de Santo Agostinho and Ipojuca (see map 5).

The Tejipió basin is not included in any of the units delimited by the state government for the RMR's small coastal rivers, despite the fact that, due to its location, it should be included in GL2 (see map 6). Its exclusion is not justified by the size of the main river or the area drained by the basin, as there is no' great discrepancy between it and those formed by the other rivers (see table 2). The characteristics of the Tejipió basin are shown below.

MAP 4: GRAPHIC REPRESENTATION OF THE BODIES OF WATER IN THE GROUP OF SMALL COASTAL RIVERS 1 - GL1.

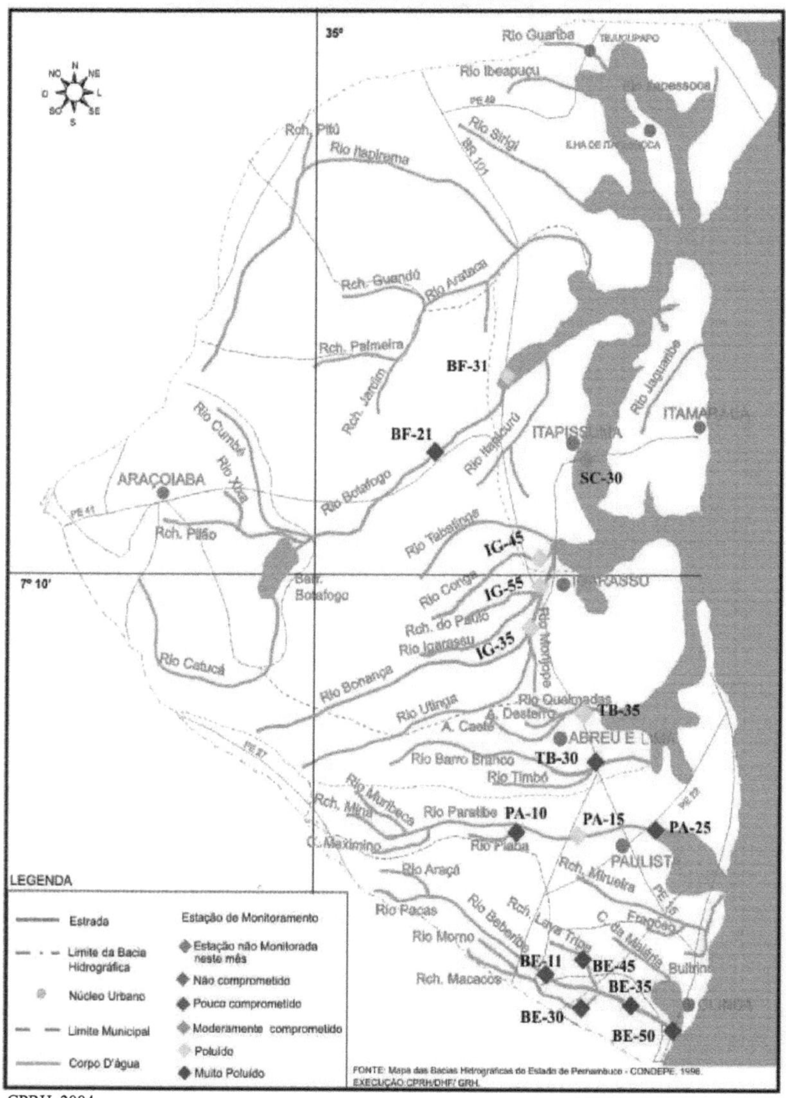

Source: CPRH, 2004.

MAP 5: GRAPHIC REPRESENTATION OF THE BODIES OF WATER IN THE GROUP OF SMALL COASTAL RIVERS 2 - GL2.

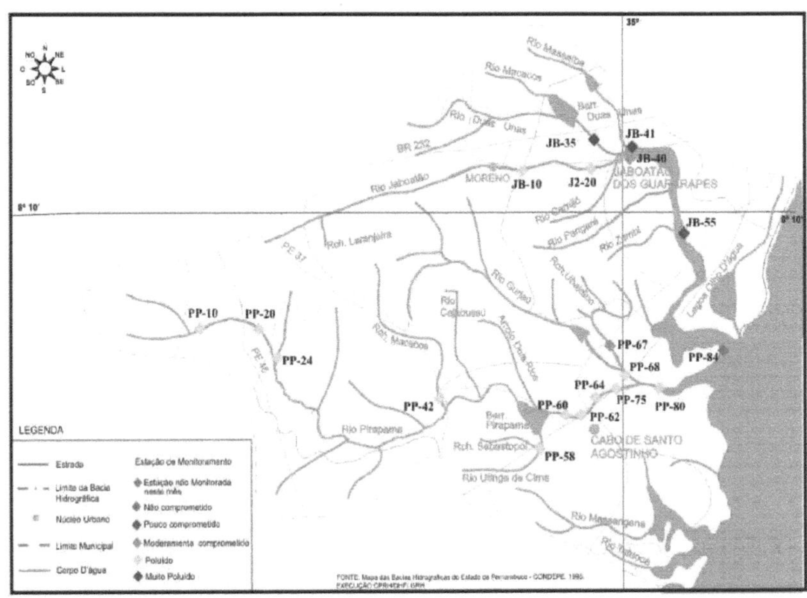

Source: CPRH, 2004.

TABLE 2: COMPARISON OF LENGTH AND DRAINAGE AREA BETWEEN THE COMPONENT RIVERS OF GL1 AND THE TEJIPIÓ RIVER.

Rivers	Extension	Basin area
Jaguaribe	9 km	18 km^2
Botafogo	21 km	476.79 km^2
Igarassu	*	143.41 km^2
Timbó	12 km	92.96 km^2
Paratibe	16 km	118 km^2
Beberibe	15 km	79 km^2
Tejipió	20.5 km	93.2 km^2

Source: CPRH, 2004.

The Tejipió river basin is a hydrographic complex made up of the Tejipió, Jiquiá and Jordao rivers (see photos 3, 4, 5 and 6), the Pina river, the Setúbal canal and other smaller watercourses, such as the 21 canals that make up its water network, distributed as follows: 5 in the Tejipió basin, 17 in the Jiquiá basin and 2 in the Jordao basin; as well as being connected to the Capibaribe river via the latter's dead bank.

Its entire length is contained within the Recife Metropolitan Region (RMR), with the main river rising in Fazenda Mamucaia, in the municipality of Sao Lourengo da Mata, and extending 20.5 km from its source to its mouth together with the Jordao and Pina rivers, in the city of Recife (see map 7). It has a total area of 93.2 Km^2 and drains portions of the municipality of Sao Lourengo da Mata (4.2 Km),^{2}
Jaboatao dos Guararapes (21.4 Km^2) and the city of Recife, which has the largest share, corresponding to 67.6 Km^2 or the equivalent of 73% of the total[126] (map 8).

Taking into account the division of Recife into Political-Administrative Regions (RPA's) (see map 9), the Tejipió and its tributaries drain all of RPA's 5 and 6, and partially RPA 4, through the Jiquiá sub-basin, whose sources are in Mata da Várzea, in the São Joao da Várzea village.

In the metropolitan context, it is the third most important basin, after the Capibaribe and Beberibe basins. However, from a drainage point of view, it is of paramount importance to the capital, since it concentrates the largest network of water resources in the city[127] (draining almost all of the urbanized area to the south and southwest) and, among the basins located within the territory of Recife, it is the one with the highest percentage of urbanized area (80%)[128] .

---

[126] RECIFE, City Hall. Department of Urban and Environmental Planning. Recife City Structuring Program - PROEST 1. *Study on the importance of the Tejipió River Basin for the Recife City Structuring Program - PROEST-1*. Recife: 1996. p. 21.

[127] According to surveys carried out for PROEST-1, the Tejipió river basin consists of 64.77 km of rivers (Tejipió, Jiquiá and Jordao); 45.52 km of canals and 239.31 km of culverts. Taking into account the length of the canals, the average presented for this is 0.69 Km/Km^2 , higher than the average for Recife, which is 0.44 Km/Km^2 . Added to the length of the rivers, the value for the basin rises to 1.67 Km/Km^2 , which is "considered good for an area with the same topographical characteristics as this area, i.e. not very hilly, with low levels and a shallow water table" (PCR:1995 apud BEZERRA: 2000, 152).

[128] RECIFE, City Hall. Op. cit. p.21.

Photo 3

Tejipió River in its middle course, in this stretch we can see the presence of secondary riparian vegetation. Author: Paulo Tavares, 2005.

Photo 4

Tejipió River in its lower course. Author: Paulo Tavares, 2004.

Photo 5

Jiquiá River. Author: Paulo Tavares, 2005.

Photo 6

Rio Jordao. Author: Paulo Tavares, 2005.

MAP 8: TEJIPIÓ RIVER BASIN IN THE CITY OF RECIFE WITH THE AREA OF INTERVENTION BY PROEST-1.

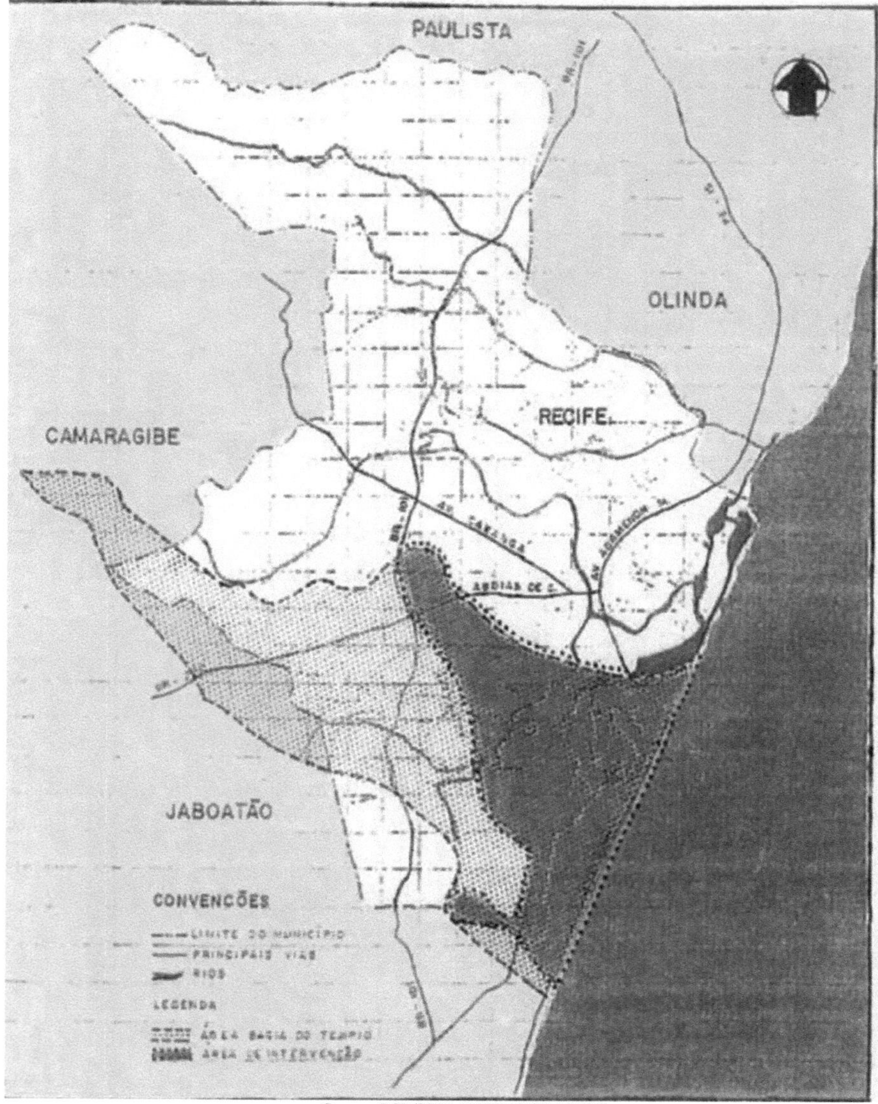

Source: PCR/SEPLAN/PROEST-1/1996

MAP 9: DIVISION OF THE CITY OF RECIFE INTO POLITICAL REGIONS ADMINISTRATIVE (RPA's).

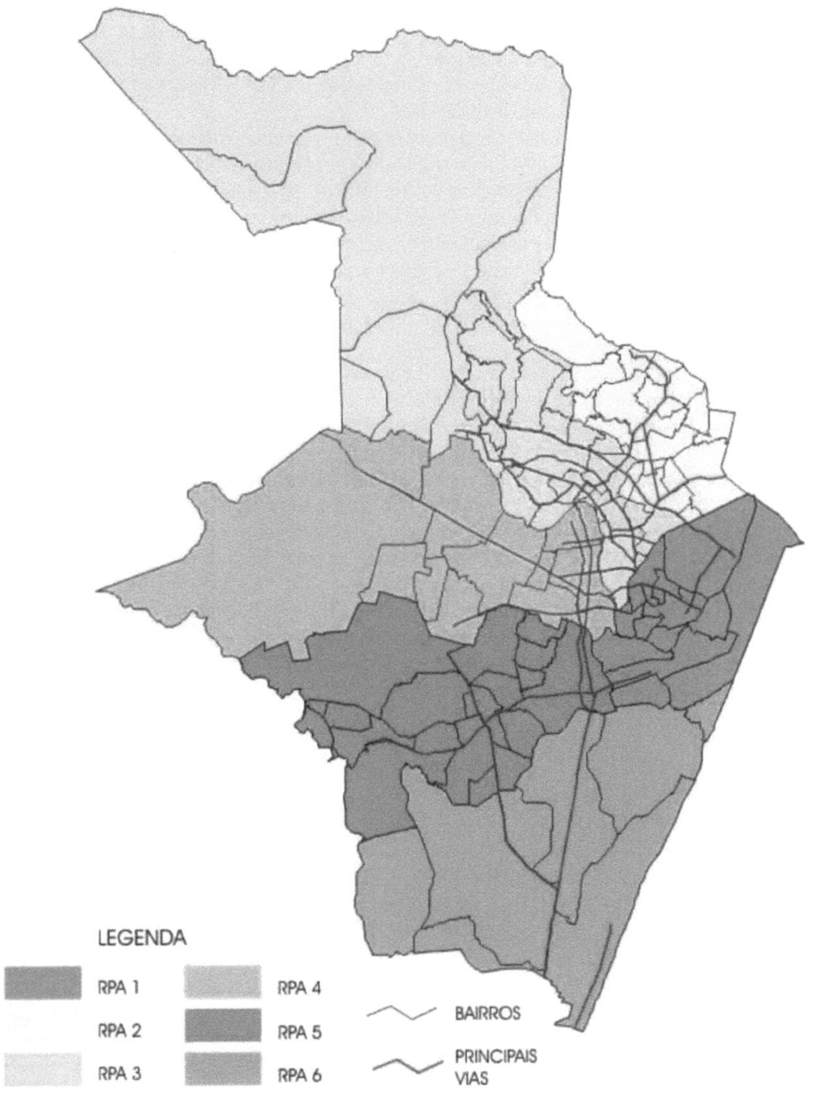

Source: PCR/DIRBAM/DEIP/Regioes Político-Administrativas do Recife/2001.

Although the Tejipió river basin is not very rugged, which means that it does not vary greatly in altitude, it is made up of two topographically distinct zones (the hills and the plain) and the type of problems encountered in it vary according to the topography. In the more elevated areas, there are processes of sliding[129] and erosion of the slopes, the latter of which, depending on the local conditions,

---

[129] A process characteristic of sandy-clay slopes, where the excessive accumulation of water in the soil causes it to become plastic and denser, causing part of it to slide down the slope.

can lead to gullies, such as those seen in the Ibura hills, which are predominantly sand-clay[130] . As a result of these processes in the higher parts, we have the silting up of the drainage elements in the shallows.

In the lowlands, where most of the Tejipió basin is located, according to the division of the city of Recife into environmental units (see map 10), the main problems are linked to drainage and are shown in the form of flooding[131] and inundations[132] , which have their causes linked to a process of urbanization that has taken place at the expense of occupying the natural space of the waters and the physical characteristics of the lowlands listed above.

Thus, even though the basin is located in a predominantly flat area with low altitudes (between 3 and 10 meters, corresponding to the Pleistocene and Holocene marine terraces), it has some rugged stretches to the west and southwest corresponding to the hills of the Zona da Mata, made up of the Crystalline Embayment, the Cretaceous deposits of the Cabo Formation and the sandy-clay deposits of the Barreiras Formation. The hills are residual reliefs that owe their dissection to various cycles of erosion that have occurred since the exposure of the forming material (mantle of alteration of crystalline and/or sedimentary rocks) to the surface.

Details of the geological-geomorphological conformation of the basin can be found in sub-chapter 2.1.

In terms of climate, the Tejipió river basin, like much of the Pernambuco coast, is under the influence of the As' climate (hot and humid climate with autumn-winter rains), according to the Koppen classification, also known as coastal "pseudo-tropical"[133] , which is responsible for high rainfall rates of between 1,800 and 2,000 mm/year and high relative humidity.

The characteristics of the climate, which is hot and humid, combined with the soil conditions, with the predominance of deep soils, due to chemical weathering of igneous rocks, have favored the predominance of natural vegetation formed by evergreen and subevergreen forests (known by the generic name of Atlantic Forest), which are characterized by their high density, wide variety of species and large size.

The forest vegetation originally covered the areas corresponding to the hills and alluvial floodplains. In the areas closest to the coast, restinga vegetation predominated and on the beaches there were coconut groves. In river estuaries and in areas susceptible to the influence of marine waters, mangroves emerged as the characteristic vegetation.

---

[130] For the gullying process to occur, in addition to the predominantly sandy composition, there must be a water table close to the surface.

[131] We consider flooding when the rise in water level does not cause the flood to overflow beyond the maximum level of the river's main channel.

[132] We consider flooding to be the phenomenon of water overflowing from the drainage channel into marginal areas (floodplain, floodplain or larger riverbed) when the flood reaches a level above the maximum level of the main river channel.

[133] ANDRADE, Gilberto Osório de. *Some Aspects of the Natural Framework of the Northeast*. Series of Regional Studies. SUDENE. 1977. p. 12-16.

MAP 10: ENVIRONMENTAL UNITS IN THE CITY OF RECIFE

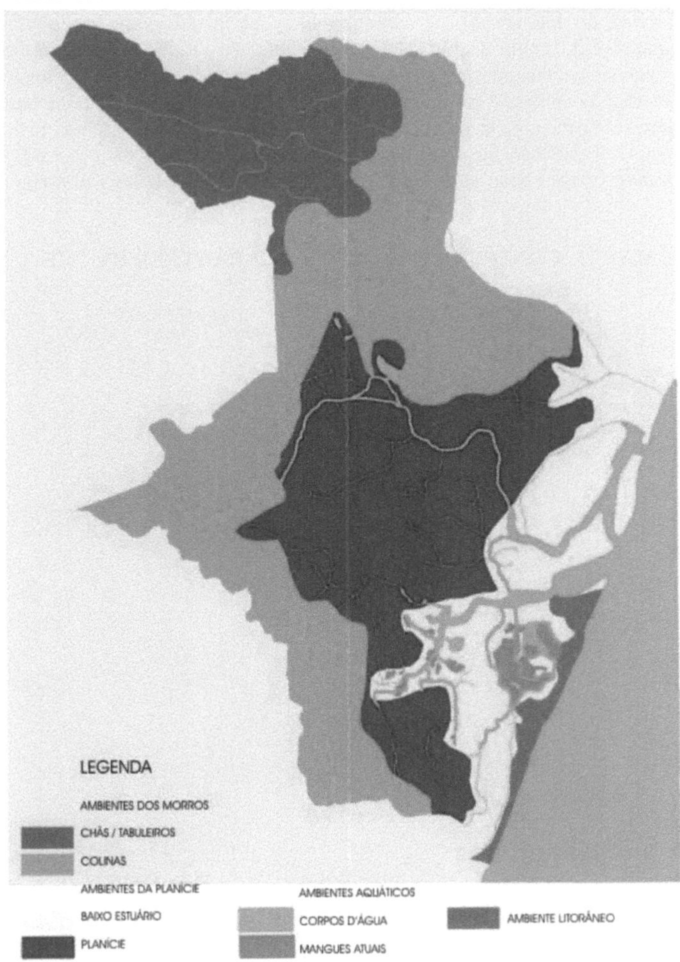

Recife City Hall: 2001.

The high degree of urbanization in the coastal areas of Pernambuco has contributed significantly to the de-characterization of forest ecosystems throughout the state. A few remnants of forest have survived this process, generally made up of dense capoeiras (secondary vegetation) in areas where the forest can recover, and remnants of original vegetation, recently transformed into conservation units[135] (ZEPA's - Special Environmental Protection Zone) by law 16.176/96, Recife's Land Use and Occupation Law, represented within the basin by the Matas do Curado, Sao Joao da Várzea, Engenho Uchoa and Jardim Botánico (see map 11); others are protected by the RMR's system for protecting water sources and ecological reserves[136] , made up of the Matas de Mamucaia, do Círculo Militar, Mussaíba and Jangadinha.

---

[135] According to the Environment and Ecological Balance Code of the City of Recife, Law No. 16.242/96, conservation units are: areas of municipal territory, including jurisdictional waters, with relevant natural characteristics, of public or private domain, legally instituted by the Public Power, with defined objectives and limits, under special management regimes, to which adequate protection guarantees are applied.

[136] State Law No. 9.980/86 and State Law No. 9.989/87.

From the above, we can say that the basin, as a discrete unit of the earth's surface, can be individualized based on the geological, geomorphological, pedological and phytogeographical characteristics that give it its main features. However, with regard to the processes developed within the basin, the analysis of the interrelationships between these elements is no longer enough to understand its internal dynamics. With the intensification of occupation and the consequent expansion of the urban sprawl over the area in question, the importance of the time variable, in addition to the geological or historical variable, has been accentuated in understanding the processes in the basin, from the perspective of the social relations of production and the relations between man and the environment.

MAP 11: CONSERVATION UNITS GUARANTEED BY THE LAND USE AND

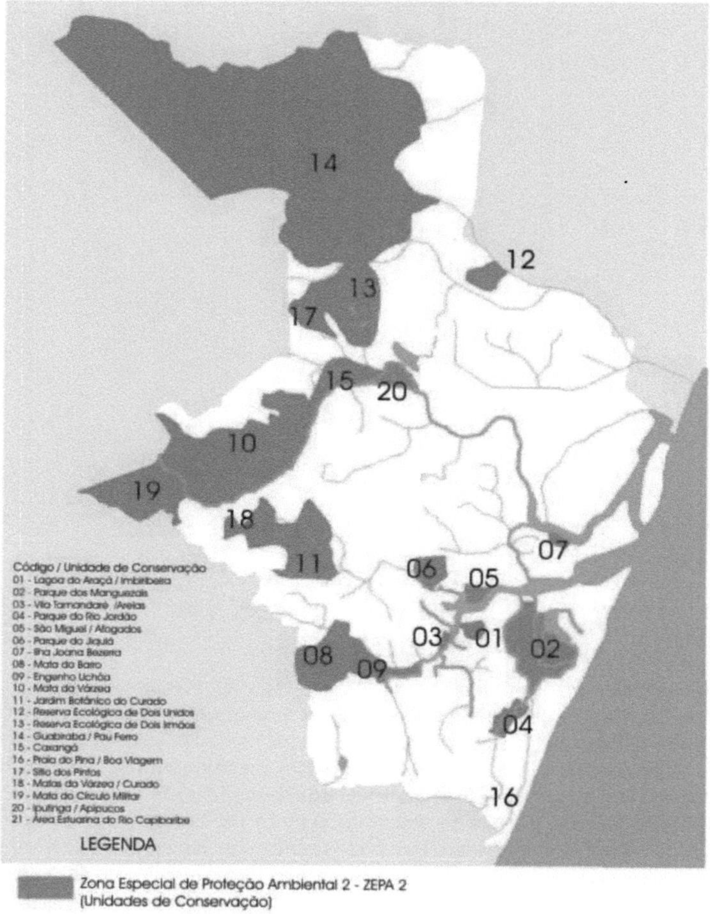

OCCUPATION LAW OF THE CITY OF RECIFE.
Source: PCR/DIRBAM/DEIP/Regioes Político-Administrativas do Recife/2001.

The alterations imposed by man on physical processes have amalgamated social dynamics with physical dynamics, so that understanding the latter necessarily involves understanding how the former is distributed in space and its nature, which will differentiate the level and intensity of impacts and, dialectically, the occurrence of natural processes will condition the distribution of human activities.

CUNHA (1998) highlights the fact that over the last three centuries "human activities have increased their influence on drainage basins"[137] . And this happens without there necessarily being any human intervention in the river channel, i.e. indirectly, as a result of society's activities in the basin (removal of vegetation, agricultural practices, urbanization) which modify the discharge behaviour of the watercourses and the supply of sediment to the river. The phenomenon explained in this way clarifies the great current interest in man as a geomorphological agent, given the awareness of his power to transform.

The transformations that have taken place as a result of the intensification of the occupation process in the Tejipió river basin will be discussed in the next sub-chapter.

3.2 Assessing the Damage: population dynamics and the socio-environmental dimension of the Tejipió River Basin.

Nature is the continent and content of man, including objects, actions, beliefs, desires and perspectives" Milton Santos

As can be seen on map 7, most of the Tejipió basin lies within the urban sphere, except for those areas occupied by ZEPAs, forest reserves and mangroves, which correspond to the yellow areas on the map. This is a result of the high rates of population growth and urbanization reported over the last 50 years; these rates appear to be a continuation of a process that began in the first decades of the 20th century.

The growing demand for land made the Tejipió basin a preferential space for occupation, as it showed the availability of land and, in some stretches, the existence of certain physical-natural elements acted to increase the value of the land, such as the coastline at Boa Viagem[138] . On the other hand, due to the heterogeneity of the surface formations presented by the basin, there were areas which, due to their natural characteristics, became unsuitable for land valuation and real estate speculation, such as flooded areas or those susceptible to periodic flooding[139] . The latter make up a significant part of the basin.

According to a study carried out by FIDEM (apud PCR: 1996, 2), Recife has a considerable liquid surface, both permanent and temporary, resulting from the combination of the physical characteristics of the fluvial-marine plain on which it is located, and the occupation made possible by landfills which, in general, did not take into account aspects relating to rainwater run-off. The study assigns the following composition to Recife's territory: 656.44 ha of permanently flooded areas; 3,306.41 ha of temporarily flooded areas; 747 ha of liquid masses (rivers, streams, canals), totaling 4,709.85 ha or the equivalent of 21.50% of the city's total surface area of 21,906.56 ha[140] .

In the study cited above, permanently flooded areas are considered to be "those occupied by mangroves and the riverside portions"[141] ; they are mainly concentrated in the lower part of the Tejipió basin, close to the estuarine zone, where the main rivers that form the basin converge. Temporarily flooded areas are more evenly distributed throughout the city and "correspond to alluvial terraces or large river channels and are more frequent on the plain"[142] .

Recife's economically and subjectively valued plains are home to a considerable portion of the urban population and the city's most valuable land. However, despite Recife's reputation as a

---

[137] CUNHA, Sandra Baptista da. *Fluvial Geomorphology.* In: GUERRA, Antonio José Teixeira & CUNHA, Sandra Baptista da (eds.). *Geomorphology: an update of bases and concepts.* Rio de Janeiro: Bertrand Brasil, 1998. p. 237.
[138] The amenities of the waterfront became an attraction for the incorporation of Boa Viagem into the city of Recife, especially from the 1920s onwards; the process of incorporation of the coastal neighborhood intensified from the 1960s onwards. The density of buildings and the complexity of the dynamics developed in this neighborhood lead BITOUN & CASTILHO (2004) to affirm the existence of a "second city" in the south zone.
[139] These areas have historically been occupied by the less privileged sectors of the population, through the construction of mocambos and stilt houses.
[140] RECIFE, City Hall. Op.cit. p. 2.
[141] RECIFE, City Hall. Op. cit. p. 2.
[142] RECIFE, City Hall. Op. cit. p. 2.

predominantly flat city, the plains occupy only 23.2% of the municipality's territory[143] . The hills occupy a larger portion and account for 67.4% of the municipality's total territory[144] .

Low-income populations predominate in the hill areas. Occupation of these areas is inadequate, amplifying and increasing the risk of material loss and loss of human life, as well as the obvious damage to the physical environment made visible by the gullies and ravines found in these areas[145] .

Occupation of the hills intensified between the 1960s and 1970s, stimulated by the floods that occurred throughout this period in the lowlands, the scarcity of land for the migrant or landless population in the flat areas and the implementation of government programs aimed at building popular housing in the Ibura hills: Cohab.

Currently, the Tejipió basin is almost completely occupied and is a densely populated area, with a population that corresponds to 49.71% of the population of Recife[146] . This percentage is mostly made up of low-income and lower-middle class people, who face serious problems linked directly or indirectly to drainage and basic sanitation. Despite the visible saturation of this space in some neighborhoods, according to the 2000 census, the geometric growth rates of the population are higher than those of the city of Recife, as well as the rate presented by the basin as a whole (see table 1).

TABLE 1: GEOMETRIC GROWTH RATE OF THE RESIDENT POPULATION IN THE COMPONENT NEIGHBORHOODS OF THE TEJIPIÓ RIVER BASIN

Neighborhood	Population (inhab.)		Tax growth (%)
	1991	2000	
Boa Viagem	89.684	100.388	11,9
Pina	26.781	27.422	2,3
Brasilia Teimosa	16.919	19.155	13,2

---

[143] RECIFE, City Hall. Recife *in numbers*. Recife: 1999.
[144] The remaining 9.3% corresponds to liquid surfaces.
[145] For more information on appropriate ways of occupying hills, see the *Guide to Hill Occupation: Recife Metropolitan Region of* the Viva o Morro Program.
[146] RECIFE, City Hall. Op. cit.

Imbiribeira	39.247	46.471	18,4
IPSEP	27.386	25.714	-6,1
Drowned	36.770	36.146	-1,6
Hose	9.159	8.734	-4,6
Mustard	12.602	12.693	-7,2
Jiquia	7.151	7.802	9,1
Estancia	10.828	8.934	-17,4
San Martin	21.682	22.959	5,8
Toast	26.661	29.510	10,6
Engenho do Meio	10.983	10.560	3,8

Jardim São Paulo	30.772	29.614	3,6
Sands	29.578	30.365	2,6
Cagote	4.755	8.427	77,2
Clay	19.828	31.111	56,9
Ibura	35.960	43.681	21,4
Cohab	49.396	69.134	39,9

According to the 2000 IBGE census, Recife has a geometric population growth rate of 12.1% for the period 1991-2000; in the same period, the Tejipió river basin has a rate of 12.9%. Taken as a whole, the population growth seen in the basin is not significant. However, when analyzed through the rates presented by neighbourhood, it can be seen that in nine of the 26 neighbourhoods in Recife that make up the basin, population growth was well above the municipal average. However, seven of them show a decrease in population, as evidenced by their negative growth rates (IPSEP -6.1; Afogados -1.6; Mangueira -4.6; Mustardinha -7.2; Estância -17.4; Coqueiral -11.6; Totó -6.8).

The neighborhoods showing the greatest growth in ascending order are: Caçote 77.2; Barro 56.9; Cohab 39.9; Curado 35.2; Sancho 22; Ibura 21.4; Várzea 19.9; Imbiribeira 18.4 and Brasília Teimosa 13.2.

The population growth observed in the neighborhoods listed above can be explained by factors such as the construction of housing estates and the increase in the number of "invasions", with the occupation of riverside areas, hillsides and the surroundings of municipal preservation units (amortization zone) (see photos 7, 8 and 9). However, the difficulty of accessing some communities and the lack of data on this growth from the City Hall are obstacles to a real understanding of the spatial distribution of this growth and its causes.

Occupation of the basin is currently taking place in two ways: in a disorderly fashion, with the low-income population occupying the few spaces available, or in an orderly fashion through real estate developments in some of its neighborhoods, such as Barro and Tejipió, which already have gated residential complexes that didn't exist a decade ago, and the construction of new objects of this nature in them.

Photo 7

Riverine occupation on the Island of God. Author: Paulo Tavares, 2004.

Photo 8

Occupation of the hills. View from Alto da Bela Vista in Coqueiral. Author: Paulo Tavares, 2005.

Photo 9

Partial view of the Jardim Botánico Residence residential complex, built near Recife's Botanical Gardens. Author: Paulo Tavares, 2005.

The land needed for these developments is obtained through the purchase of old farmhouses and manor houses, in some cases the old buildings are given new functions, such as a party room or condominium administration (see photos 10 and 11). In this process, the green areas, preserved or, as is often the case, cultivated by the former owners of the land[146] , become an element of the building's value.

View of the Casa Solar Residence in Barro. In the detail, you can see completed and inhabited buildings and buildings under construction, one of which can be seen in the larger photo. Author: Paulo Tavares, 2005. **photos 10 and 11**

[1] In this case, most of them are fruit trees, but there are some rare species that aroused the interest of the municipal authorities in their preservation.

Throughout the 1960s and 1970s, an important agent in the production of this space was the state, through investments in the construction of villas and cohabs, such as Vila Cardeal e Silva, Vila das Lavadeiras, Vila IPSEP or housing estates such as Ignês Adreazza.

When occupation is promoted by the low-income population, it takes the form of "invasions" and popular allotments. This process of continuous occupation of available spaces gives rise to what are now known as "communities", such as Mangue Seco and Miguel Arraes, both in the Areias neighborhood, which arose from the recent landfill (just over three years, according to local reports[147]) of the mangrove swamp on the banks of the Tejipió River near Avenida Recife (see photos 12, 13 and 14).

Due to the spontaneous way in which they arise and the lack of infrastructure works to minimize environmental impacts (or any kind), construction generally causes serious damage to the region's delicate drainage system, which is already deficient due to its own physical characteristics (low altitudes and, due to low and poorly made landfills, no steep gradients; in addition to having a shallow water table). Also during the process, there is an increase in sanitary waste pollution as a result of this dynamic not being accompanied by a restructuring of the sewage system, which often doesn't serve the surrounding areas either[148] .

[147] Information obtained from field research carried out on March 23, 2004.
[148] First of all, it is necessary to put an end to the myth that 25% of the city is sanitized. The fact that the majority of the population in neighborhoods such as Boa Viagem are able to evacuate their sewage does not mean that it is sanitized, as part of the connections are made through the drainage network, releasing the effluents without proper treatment into the city's water bodies (BITOUN, 2002).

Photo 12

I Occupation of mangroves in the Community of Miguel Arraes, Areias. Author: Paulo Tavares, 2004.

Photo 13

Recent occupation of the banks of the Tejipió River in Mangue Seco, Areias. Author: Paulo Tavares, 2004.

Photo 14

Occupation of the banks of the Tejipió in the Totó neighborhood. It does not differ greatly from the conditions found in areas of recent occupation. Author: Paulo Tavares, 2003.

If we take into account the spatialization of occupation typology and the types that are located near environmentally degraded areas, socio-spatial relations and the distribution of the class structure are of crucial importance in understanding the environmental problem[149] . This is because the different social classes exert different pressures on the state apparatus to demand public policies aimed at providing a given area with the infrastructure needed to improve its environmental conditions, the design and implementation of which reflects on the improvement of the population's living conditions[150] ; and they will also have varying conditions of location in the urban space depending on income variations. In this way, the less privileged social classes will be relegated to environmentally unstable areas (hillsides, wetlands).

The problem analyzed in this way prevents the victims of environmental impacts (low-income communities) from being seen as the culprits.

As we pointed out earlier, the causes of pollution of a river, which by extension can be applied to the pollution of any other natural element as well as the environment itself, can vary in time and space due to the feedback mechanism and variations in the interaction of the component elements of this systemic totality. In this way, the causes of environmental impacts are not always to be found in the low-income communities that experience their consequences most violently and permanently.

The lack of a sewage system, combined with the idea that a small percentage of the city is sanitized, means that a portion of the population is forced to live directly with the waste they produce and that produced by the houses and buildings in the so-called sanitized areas, Because of this concept, some sectors of society become complacent about the lack of sanitation, since they are not directly affected by its main effects, and do not demand the implementation of this service from government officials who, in turn, do not see a political return on investments in underground infrastructure. In short, "the lack of sanitation (...) affects the unequal unequally"[151] .

In addition to social impacts, there are environmental impacts. The degradation of rivers located in urban areas, as well as mangroves, as a result of domestic waste is visible and can be measured using indices such as BOD, DO and fecal coliforms (FC), among others. The growing volume of sanitary effluents over the decades has led to a stage of degradation of water resources that is difficult to recover from, where at certain times of the year rivers have dissolved oxygen levels equal to zero[152] .

Aiming at the environmental recovery of the Tejipió, Jiquiá, Pina and Jordâo river basins, in 1995 the City Council launched the Recife City Structuring Program - Phase 1 (PROEST 1), also called the Tejipió River Basin Urban and Environmental Recovery Program, which emerged as the materialization of one of the urbanization programs previously proposed by the Recife City Development Master Plan (PDCR)[153] .

The objectives of PROEST 1 included the enhancement and integration of environmental resources for the use of the population in a planned manner based on a new model of urbanization, based on the environmental sanitation of the lower reaches of the basins mentioned in the previous paragraph, through the provision of infrastructures for this purpose (sewage treatment plants, pumping stations, among others).

---

[149] COELHO, Maria Célia Nunes. Op. cit. p. 20.

[150] In Recife's social environment, marked by unequal cultural and economic conditions, people's experiences of environmental problems are so diverse that they make it difficult to mobilize the population more broadly in the search to solve them.

[151] BITOUN, Jan. *Sanitation in Recife: how broadening the debate can tackle the crisis*. In: RECIFE: City Hall. *How to clean up Recife as quickly as possible: sanitation for a better life*. Recife: Recife City Hall, 2002.

[152] For the specific case of the Tejipió River, see table 1 on page 23. For more information on other rivers, see the CPRH publications on monitoring the environmental quality of the waters in the Pernambuco state river basins that make up the PQA.

[153] "The PROEST - 1 area involves, totally or partially, 23 neighborhoods in the city, which together occupy an area of 6,596 ha (30.7% of the city) and were home to a population of 503,115 inhabitants in 1991 (38.8% of Recife's total), including Boa Viagem and Imbiribeira, which are among the most prominent from an economic point of view. These figures correspond to an average density of occupation of 76.27 inhabitants per hectare, higher than the average for Recife, which is 60.35 inhabitants per hectare" (PCR: 1996, 26).

The actions of PROEST 1 would focus on the Tejipió basin[154] because it is one of the most dynamic areas of the city[155], because it is subject to strong pressures on the environment (disorderly land occupation, popular settlements on the banks of rivers and canals, pollution of water bodies by sewage, garbage and others) and because in recent years it has been the target of little investment by the public authorities. These actions comprised the following sub-programs: sanitation (construction of networks, trunk collectors, outfalls, pumping stations and treatment units), urban cleaning (expansion and implementation of selective waste collection), urbanization (recovering and urbanizing the banks of the Pina and Ilha de Deus mangroves by setting up physical infrastructure and reorganizing the low-income settlements that occupy their banks) and environmental education (setting up an environmental education program in the PROEST - 1 area)[156].

Thus, the project would be justified by the main environmental problems affecting the basin, such as: water and soil pollution as a result of sewage and garbage being dumped in rivers and canals; frequent flooding of the streets, related to the insufficient drainage system, silting up and obstruction of galleries and riverbeds; destruction of the Pina Mangrove by water pollution and landfill as a result of the expansion of occupation on its banks due to pressure from private real estate agents and, above all, the low-income population.

In addition to the restoration of degraded areas and their environmental sanitation, the infrastructure that would be promoted by PROEST 1 in the lower estuary of the Tejipió basin was intended to attract new economic investment to the area, especially to the more dynamic neighborhoods such as Boa Viagem, Pina, Imbiribeira and Afogados. This point, regarding the attraction of new investments, is presented in a developmentalist discourse, stating the importance of the expected investments coming to the city and emphasizing their role in generating jobs and income.

However, the project makes no mention of the real estate development that would result from such an investment by the public authorities in the provision of infrastructure in this area, which would complete the process of increasing the value of the areas bordering the banks of the Jordao River, which began in the 1970s and 1980s with the installation of the Recife Shopping Center and the implementation of projects such as CURA[157] and Nassau[158], including the areas that were uncovered and environmentally degraded, such as the Pina Mangrove, Ilha de Deus[159] and the southernmost stretch of the Jordao River; implying structural (or structuring) interventions and the removal of favelas[160].

In this way, PROEST 1 would lead to an increase in real estate speculation in the area mentioned in the previous paragraph, due to the increase in the differential rent II obtained by owning the land located there (which includes the neighborhoods of Imbiribeira - the eastern sector along the Jordao River -, Boa Viagem, Setúbal and Pina - the western sector), making it impossible, in theory, to maintain the homes of the low-income population in these areas.

It is necessary here to differentiate between differential income I and II, obtained through the ownership of urban land.

According to HUARACHI (1982), the differential rent I corresponds to the location and accessibility aspects of the land, with location being determined by the distance between the site of

---

[154] See map 8 on page 64 for a visualization of PROEST's direct intervention area in the Tejipió basin.
[155] In making this statement, Recife City Hall has certainly taken into account the dynamism of Boa Viagem and the neighborhoods around it.
[156] RECIFE, City Hall. *Study on the importance of the Tejipió River Basin for the Recife City Structuring Program - PROEST-1*. Recife: 1996.
[157] The CURA Project (Comunidade Urbana para Recuperado Acelerada - Urban Community for Accelerated Recovery), created in 1971, set out "to give better order to urban land, improve infrastructure and correct the problems of real estate speculation" (HUARACHI: 1982, 28).
[158] The Nassau Project of Recife City Hall was developed with the aim of solving the problem of flooding by lining and widening some of the city's canals and urbanizing their banks.
[159] The community was called "Ilha sem Deus" (Island without God) before the construction of the bridge that gives it access to Imbiribeira, because it was said that it was completely forgotten by the government and by God.
[160] Within the program, there were projects exclusively aimed at studying the reality of the Pina Mangrove, Ilha de Deus and the Jordao River in order to promote their urbanization, given their importance for PROEST and as an area of urban expansion for the city.

the land and the place where urban facilities and urban functions are located (place of work, place of commerce, leisure area). Other factors that influence the value of land are: i) the function for which it will be used and that of the area in which it is located; ii) aspects of location related to construction differences and accessibility[161] .

The area in question, near the Recife Shopping Center, has a privileged location in terms of producing this type of differential income, type I, but this potential is not converted into income due to the low attractiveness caused by the precariousness of the installed infrastructure and the degradation of the surrounding environment. The same applies to Afogados.

According to HUARACHI (1982), differential rent II is generated from investments made in the area, which can be made by the state and/or private individuals[162] . This type of rent also takes into account the Land Use and Occupation Law, which determines the maximum level of investment to be made in the area and its surroundings.

There was a specific project in PROEST 1 aimed at the urbanization of the Pina and Jordao river basins which, in a veiled way, would continue the improvements previously implemented by the Nassau Project, expanding the space towards
Setúbal[163] . In 1995 and 1996, the City Council lined and straightened the Jordao River in the stretch between Barao de Souza Leao and the REFFSA bridge, since it had not been covered by Nassau. With PROEST 1, the aim was to solve the problem of flooding and the strangulation of the riverbed in some stretches by lining and channeling the Jordao River to its source.

In the other locations in the PROEST 1 direct intervention area, the planned works were focused on the installation of pumping stations, ETEs, trunk collectors, among others; aimed at minimizing the damage to the environment caused by the absence of a sewage collection and treatment system.

If, on the one hand, the implementation of PROEST 1 would possibly trigger the resumption of the urban renewal process[164] , with the removal of the low-income population and the materialization of new investments in the lower estuary of the basin, on the other hand, its non-execution has maintained the previous conditions of insalubrity and deterioration of the social and physical environment.

And this situation of poor quality of life and lack of infrastructure has worsened over the years, with PROEST 1 being the last major project aimed at environmental sanitation and the urban (re)structuring of the Tejipió basin.

The lack of studies aimed at assessing the basin's carrying capacity and the increase in the population occupying it further aggravates the degradation of natural elements, since this necessarily implies, in view of the city's sanitary conditions, an increase in the amount of domestic effluent discharged into water bodies and, consequently, an increase in eutrophication levels, as well as the pressure of construction on the basin.

Waterproofing, resulting from the process of urbanization and the densification of buildings, reduces the amount of water that infiltrates to replenish aquifers and to form sub-surface currents; it therefore reduces the flow of rivers during the dry season, even causing some smaller watercourses to disappear[165] . The reduction in flow causes a reduction in the solubility capacity of the water, increasing pollution levels and interfering with the environmental quality of the waters and thus the

---

[161] "With regard to accessibility, this is basically determined by the circulation function. Land located in areas well served by road and transportation systems achieves higher values" (HUARACHI: 1982, 14).

[162] "The investments made by the private sector take the form of the construction of medium and high-end buildings in areas where the city is expanding. It should also be noted that these investments contribute to the increase in value of urban land" (HUARACHI: 1982, 23).

[163] "With the infrastructure works, conducted under the pretext of drainage, spates of expansion are created" (GOMES: 1997, 239).

[164] According to HUARACHI (1982), "the term urban renewal refers to the process of removing slum areas and replacing the spaces with a spatial occupation with different functions or even a housing function, with the constructions being of a medium or high building standard". The process of urban renewal has been taking place in Boa Viagem in particular since the middle of the 20th century.

[165] For more information on the implications and risks of waterproofing, see TUCCI 1995 & 2002.

conditions for the reproduction of animal and plant life.

The fact that a considerable proportion of the population lives close to riverbeds, streams and canals and that many diseases are transmitted by waterborne means that the issue of water degradation is a pertinent topic for discussion, particularly in a city with the physical characteristics of Recife.

The lack of government planning that is coherent with the socio-economic-environmental realities of the Tejipió river basin and the dynamics of the physical environment that underpins it, makes it increasingly difficult to solve the problems encountered in the different localities that make up the basin. Delaying solutions to problems that should be taken now due to their urgency.

FINAL CONSIDERATIONS

Throughout the dissertation, it has been possible to verify, within the complex context in which phenomena occur, the impossibility of apprehending reality in a fragmented way, where the analysis of physical dynamics can be dissociated from the transforming action of man and, likewise, considering socio-spatial dynamics free from the conditioning of the surrounding environment; which characterizes a reductionist view of the world that allows environmental impacts to be considered as products of the Society/Nature interface and not as a dynamic process, as defended in this work.

The overlapping of objects and human actions in the environment establishes an interrelationship based on a system of exchanges that does not occur directly, with the establishment of a rigid relationship between cause and effect, because, depending on the particular properties presented by each of the subsystems that make up the system in question, the same external influence can cause different results; therefore, it is necessary to assess each of the phenomena within its own context, where the probabilistic correlation of their effects can be taken into account. In this way, as we have tried to demonstrate in this study, General Systems Theory, embodied in Complexity Theory, is an important paradigm for understanding the elements that make up reality in their competition, antagonism and complementarity.

In the context of the integrated study of socio-environmental dynamics, river basins are extremely important because they are complex systems that include within their perimeters a series of processes, both social and physical, which are constantly changing. The drainage network is the most sensitive element to the changes seen in the basin, since most of the outputs of the other subsystems go to the drainage network, and it is the main point where the dynamic interactions of the anthroposociological whole converge with those developed by the river basin as a whole.

Following the line of reasoning set out in the previous paragraphs, it became clear that the current state of the Tejipió river basin is the result of the process of occupation of Recife, which, in the area drained by it, took place spontaneously, intensifying during the 20th century, meaning that it did not take into account the physical characteristics presented by the different spaces occupied in the basin, acting to destabilize the existing environmental systems.

On the plain, occupation has accentuated the drainage problems naturally present due to its low altitude and shallow water table, as they have been caused by low and poorly made embankments made by the population itself. As these embankments do not have a steep gradient, they make it easier for water to remain in this system for longer, favoring the occurrence of floods. In addition to the problems on the plain, there is a lack of basic sanitation, increasing the possibility of contamination by water-borne diseases and reducing the quality of life in the riverside and flooded areas, which are subject to frequent flooding.

On the hillsides, human intervention, by changing the way the slopes are balanced in order to build houses, by cutting into the barrier and removing the vegetation cover, intensifies existing erosion processes or provides the conditions for them to take place. As erosion processes are conditioned by the granulometric composition of the material to be eroded, they will vary in the Tejipió basin. When they occur on predominantly clayey sedimentary packages, they take the form of sliding barriers, and when they occur on predominantly sandy packages, they take the form of ravines and gullies excavated by the action of linear erosion on these surfaces.

Erosion increases the risk of material loss or loss of human life on the hills. In the western portion of the Tejipió basin, there are several risk areas which, because they are at an advanced stage of degradation, require structural interventions such as the construction of retaining walls; in others, however, the planting of grasses is enough, a solution that has been little implemented by the government in favor of structural interventions.

The material carried from the higher areas accumulates in the shallows and riverbeds, silting them up and aggravating the occurrence of floods during the rainy season.

Another aggravating factor for drainage in the basin is the fact that the drainage network is drowned in many stretches, either by water from the groundwater, which is influenced by the tide, or by clandestine sewage connections that deposit domestic effluent in the network intended for rainwater runoff, resulting in reduced transport capacity during the wet season and overflowing galleries, flooding streets and avenues.

Solid waste, accumulated in the beds and on the banks of rivers and canals, also contributes to the galleries working under pressure. The difficulty of access to a regular collection system in some places is an excuse for many people to dump their waste in unsuitable areas. For this reason, we recommend implementing the environmental education programs provided for in PROEST 1 as a way of raising awareness, with a view to reducing the volume of solid waste in the galleries, rivers and canals.

During the research, it was difficult to obtain specific data and information on the Tejipió river basin, since it is not included in the government programs dedicated to the state's river basins; a fact that cannot be justified by the size of the basin or its length. The only project focused on it was PROEST, but this has a limited amount of information.

The complexity of the phenomena covered in this dissertation makes it difficult to draw up a prognosis with definitive solutions to the problems encountered along the Tejipió river basin, as the dynamic nature of the elements involved in the processes developed in the basin makes it difficult to draw definitive conclusions or construct precise future scenarios. However, it is clear that there is an imminent need for interventions aimed at the environmental and urban recovery of the basin, through the provision of infrastructures that guarantee a reduction in the impacts caused by urbanization and measures that ensure the recovery of the basin's ecosystems.

BIBLIOGRAPHICAL REFERENCES

ALHEIROS, Margareth et. all. *Geological Map of the City of Recife.* SEPLAN/Prefeitura da Cidade do Recife. 1995.

. *Landslide Risks in the Metropolitan Region of Recife.* UFBA: Postgraduate Course in Geology. Doctoral Thesis, 1998, 135pp.

ANDRADE, Gilberto Osório de. *Some Aspects of the Natural Framework of the Northeast.* Series of Regional Studies. SUDENE. 1977.

BARROS FILHO. Mauro Normando Macedo. *The Specificities of Urban Socio-Spatial Heterogeneity. The case of the Torrees ZEIS in the city of Recife.* Master's dissertation. Recife: MDU, 2000.

BITOUN, Jan. *Sanitation in Recife: how broadening the debate can tackle the crisis.* In: RECIFE: City Hall. *How to clean up Recife as quickly as possible: sanitation for a better life.* Recife: Recife City Hall, 2002.

BITOUN, Jan & CASTILHO, Cláudio Jorge Moura de. *The Economic Activities of Recife: analysis of their distribution in the territory.* Recife: PCR, 2004.

BERTALANFFY, Ludwig von. General Systems Theory. Petrópolis: Vozes, 1977.

BEZERRA, Onilda Gomes. *The Cultural Representation of a Landscape: the Pina mangrove swamp and the urbanization of the south/southeast zone of Recife.* Master's dissertation project. Recife: CMG, 1998.

BOTELHO, Rosangela Garrido Machado & SILVA, Antonio Soares da. *Watershed and Environmental Quality.* In: VITTE, Antonio Carlos & GUERRA, Antonio José Teixeira (org.). *Reflexees sobre a Geografia Física no Brasil.* Rio de Janeiro: Bertrand Brasil, 2004.

BRITO NEVES, Benjamin Bley de. *The Geological Map of the Eastern Northeast of Brazil, scale 1:100.000.* USP: Institute of Geosciences. Thesis, 1983, 177pp.

CASTRO, Josué de. *Ensaios* de *Geografia Humana.* Sao Paulo: Brasiliense, 1966.
. *Men and Crabs.* Sao Paulo: Brasiliense, 1968.

CHORLEY, Richard & HAGGETT, Peter. *Physical and Information Models in Geography.* Sao Paulo: Edusp, 1975.

CHRISTOFOLETTI, Antonio. *Systems Analysis in Geography.* Sao Paulo: HUCITEC, 1979.

Geomorphology. Sao Paulo: Edgard Blücher, 1999.

Environmental Systems Modeling. Sao Paulo: Edgard Blücher, 1999.

CHRISTOFOLETTI, Anderson Luís Hebling. *Dynamic Systems: Approaches to Chaos Theory and Fractal Geometry in Geography.* In: VITTE, Antonio Carlos & GUERRA, Antonio José Teixeira (org.). *Reflexöes sobre a Geografia Física no Brasil.* Rio de Janeiro: Bertrand Brasil, 2004.

COELHO, Maria Célia Nunes. *Environmental Impacts in Urban Areas - Theories, Concepts and Research Methods.* In: GUERRA, Antonio José Texeira & CUNHA, Sandra Batista da (eds.). *Urban Environmental Impacts in Brazil.* Rio de Janeiro: Bertrand Brasil, 2001.

CONTI, José Bueno. *Physical Geography and Society-Nature Relations in the Tropical World.* In: CARLOS, Ana Fani Alessandri. *New Paths in Geography.* Sao Paulo: Contexto, 1999.

COSTA, Eda Maranhao Pessoa da. *Urban Expansion and Spatial Organization.* Recife: UFPE, 1982.

COSTA, José de Araújo da. *Environmental problems in the Tejipió River.* Monograph. Recife, UFPE: 1997.

CORRÉA, Roberto Lobato. *Geographical Trajectories.* Rio de Janeiro: Bertrand Brasil, 1997.

COUTINHO, Roberto Quental et al. *Climatic, Geological, Geomorphological and Geotechnical Characteristics of the Dois Irmao Ecological Reserve.* In: MACHADO,

Isabel Cristina et. al. *Reserva Ecológica de Dois Irmaos: Estudos em um Remanescente de Mata Atlántica em Área Urbana (Recife - Pernambuco - Brazil)*. Recife: Editora Universitária da UFPE, 1998.

CPRH. *Monitoring the Quality of Water in the Hydrographic Basins of the State of Pernambuco - 1995*. Recife: CPRH, 1995.

. *Monitoring Water Quality in the Hydrographic Basins of the State of Pernambuco - 1996*. Recife: CPRH, 1996.

. *Monitoring Water Quality in the Hydrographic Basins of the State of Pernambuco - 1997*. Recife: CPRH, 1997.

. *Monitoring Water Quality in the Hydrographic Basins of the State of Pernambuco - 1998*. Recife: CPRH, 1998.

CUNHA, Sandra Baptista da. *Fluvial Geomorphology*. In: GUERRA, Antonio José Teixeira & CUNHA, Sandra Baptista da (eds.). *Geomorphology: an update of bases and concepts*. Rio de Janeiro: Bertrand Brasil, 1998.

CUNHA, Sandra Batista da & GUERRA, Antônio José Texeira (eds.). Geomorphology of Brazil. Rio de Janeiro: Bertrand Brasil, 2001.

DOMINGUEZ, José Maria Landin et al. *Geology of the Coastal Quaternary of the State of Pernambuco*. Revista Brasileira de Geociências, vol. 20, n° 4, p.208-215.

ESTEVES, Cláudio Jesus de Oliveira. *Tourism and Water Quality on Ilha do Mel (Paraná Coast)*. Master's dissertation. Curitiba: CPGG, 2004.

FIDEM. Tejipió Basin. Recife: 1982.

GUERRA, Antonio Texeira & CUNHA, Sandra Batista da (eds.). *Geomorphology: an update of bases and concepts*. Rio de Janeiro: Bertrand Brasil, 1998.

GOMES, Edvania Torres Aguiar. *Landscape Cut-outs in the City of Recife: a geographical approach*. Doctoral thesis. Sao Paulo: USP/CDG, 1997.

HUARACHI, David Gilberto Cáceres. *Housing Policy and Urban Renewal in the Metropolitan Region of Recife*. Master's dissertation. Recife: MDU, 1982.

LIMA FILHO, Mario Ferreira de et. al. *Origin of the Recife Plain*. In: Estudos Geológicos UFPE/DEGEO (Series B - Estudos e pesquisas). Revisao Geológica da Faixa Costeira de Pernambuco, Paraíba e Rio Grande do Norte. UFPE: Recife, 1991. vol. 10, p. 157-176.

LINS, Rachel Caldas. *Some Original Aspects of the Site of Recife*. In: ANDRADE, Manuel Correia de (org.) *Capitulo* de *Geografia do Nordeste*. Recife: International Geographical Union - National Commission of Brazil, 1978. p. 81-84.

MELO, Mário Lacerda de. *Metropohzacao e Subdesenvolvimento: o caso do Recife*. Recife: UFPE, 1978.

MORIN, Edgar. *Science with conscience*. Rio de Janeiro: Bertrand Brasil, 1999.

The Method. Lisbon: Publicares Europa-América, 1997.

The Enigma of Man: Towards a New Anthropology. Rio de Janeiro: Zahar Editores, 1979.

MONTEIRO, Carlos Augusto de Figueiredo. *Geosistemas: the history of a search*. Sao Paulo: Contexto, 2000.

MOTA, Ana Cláudia de Souza. *Urban Mining in the Municipalities of Recife and Jaboatao dos Guararapes*. Master's dissertation. Recife: PPGG, 2002.

PERREIRA DA COSTA, Francisco Augusto. *Around Recife*. Recife, Fundaçao de Cultura Cidade do Recife: 1981.

REBOUQAS, Aldo da Cunha et al. *Águas Doces do Brasil*. Sao Paulo: Escrituras, 2002.

RECIFE, City Hall. Secretary for Urban and Environmental Planning. Structural Program for the City of Recife - PROEST 1. *Environmental Pre-Feasibility Study for PROEST-1 (Preliminary Version)*. Recife: 1996.

. *Study on the importance of the Tejipió River Basin for the Recife City Structuring Program - PROEST-1*. Recife: 1996.

Study of water quality in the Tejipió River Basin. Final Report. Recife, 1996.

Urban and Environmental Recovery Program. Urban Cleaning Subprogram. Recife, 1996.

Drainage System. Recife, 1996.

Dynamism of the area and the Neighborhoods of the Structuring Program. Recife, 1996.

Environmental Recovery of the Tejipió River Basin. Current characterization of surface water hydrology and hydrodynamics. Technical Report. Recife, 1996.

Water quality module for the Tejipió River Basin. Recife, 1996.

Political-Administrative Regions of Recife - Southwest Region - RPA-5 (Preliminary Version). Recife: 2001.

Political-Administrative Regions of Recife - Southern Region - RPA-6 (Preliminary Version). Recife: 2001.

Political-Administrative Regions of Recife - West Region - RPA-4 (Preliminary Version). Recife: 2001.

Municipal Profile-History and Urban Evolution. Recife: Directorate of Urban Planning, 1989.

Recife in numbers. Recife: 1999.

RIBEIRO, Luiz Cesar de Queiroz. *From Tenements to Closed Condominiums: The Forms of Housing Production in the City of Rio de Janeiro.* Rio de Janeiro: Civilizaçao Brasileira, 1997.

SANTOS, Milton. *Space & Method.* Sao Paulo: Nobel, 1997.

. *The Nature of Space.* Sao Paulo: Edusp, 2002.

SUGUIO, Kenitiro et al. *Relative sea level fluctuations during the upper Quaternary along the Brazilian coast and their implications for coastal sedimentation.* Revista Brasileira de Geociencias, vol. 15, n° 4, p.273-286.

TUCCI, Carlos E. M. et. al. (eds). *Drenagem Urbana.* Porto Alegre: ABRH/UFRGS, 1995.

VASCONCELOS, Ronald F.A. *Saneamento do Recife.* Cadernos do Meio Ambiente do Recife, v.1, n.2, Jul./Dec.. Recife: PCR, 1998.

I want morebooks!

Buy your books fast and straightforward online - at one of world's fastest growing online book stores! Environmentally sound due to Print-on-Demand technologies.

Buy your books online at
www.morebooks.shop

Kaufen Sie Ihre Bücher schnell und unkompliziert online – auf einer der am schnellsten wachsenden Buchhandelsplattformen weltweit! Dank Print-On-Demand umwelt- und ressourcenschonend produzi ert.

Bücher schneller online kaufen
www.morebooks.shop

MIX
Papier aus verantwortungsvollen Quellen
Paper from responsible sources
FSC
www.fsc.org
FSC® C105338